SpringerBriefs in Business

For further volumes:
http://www.springer.com/series/8860

Sik Sumaedi · I Gede Mahatma Yuda Bakti
Nidya Judhi Astrini · Tri Rakhmawati
Tri Widianti · Medi Yarmen

Public Transport Passengers' Behavioural Intentions

Paratransit in Jabodetabek–Indonesia

Springer

Sik Sumaedi
I Gede Mahatma Yuda Bakti
Nidya Judhi Astrini
Tri Rakhmawati
Tri Widianti
Medi Yarmen
Indonesian Institute of Sciences
Tangerang Selatan
Indonesia

ISSN 2191-5482 ISSN 2191-5490 (electronic)
ISBN 978-981-4585-23-1 ISBN 978-981-4585-24-8 (eBook)
DOI 10.1007/978-981-4585-24-8
Springer Singapore Heidelberg New York Dordrecht London

Library of Congress Control Number: 2013957993

Printed on acid-free paper

Springer is part of Springer Science+Business Media (www.springer.com)

Preface

Not unlike other big cities in the world, the Greater Area of Jakarta (Jabodetabek) has its fair share of traffic congestion. The financial loss stemmed from it was more than 5 billion Rupiah and at some point, it even had cost the people beyond the count of money. Psychological burden and environmental damages were among the vast effects brought by traffic congestion. Peak hours were frustrating hours for drivers and also the hours of higher carbon dioxide concentration for our planet. The problems associated with traffic congestions could not be underestimated anymore.

These problems had called a sense of urgency for the governments to come up with applicable solutions; bus rapid transit, monorails, overpasses and automatic tollbooth. They worked very well on some countries and failed on the others. For Jakarta, somehow traffic congestions did not cease to exist no matter how many technical solutions implemented by the government. TransJakarta has been rolling on the roads and the electric trains had been upgraded, but still, the road capacity was ridiculously outnumbered by the amount of private vehicles. This phenomenon has led us to believe that there was something more than technical and rational reasons that affect the behaviours of passengers.

Previous studies saw passengers' behaviour as something that was rationally premeditated; decisions that were rationally taken using all information available and excellent cognitive evaluation skill. This study saw it from a different point of view. It conferred about the relationships among perceived sacrifice, value, service quality, trust, image and behavioural intentions. In other words, it considered the other sides of the passengers; their affective nature and hedonic orientation.

The general purposes of this study was to create a model that could explain passengers' behavioural intention in Jabodetabek; things that were important for them, things that could make them tick and decide to use certain kind of transportation mode. The knowledge of passengers' behaviour was expected to give insight into how to solve traffic congestion in a more comprehensive manner.

This study was conducted in Jabodetabek and samples were taken from pools of paratransit passengers. The data collected by using questionnaire were then used to measure seven variables, namely image, perceived sacrifice, service quality, perceived value, satisfaction, trust and behavioural intention and their respected relationships. All measures and indicators were taken from the relevant literatures. The data analysis utilised Structural Equation Modelling under LISREL 8.80 to test the hypotheses.

The result of this study was able to discern the nature of paratransit passengers' behavioural intention. We hoped that this finding will be useful for future researches on consumer behaviour as well as policy makers in the field of transportation. Although it might not offer the whole solution, it put forward complementary answers to disentangle traffic congestion in the Greater Area of Jakarta.

Acknowledgments

We would like to thank the Indonesian Ministry of Research and Technology for supporting our research. We would also like to thank Ika Fitriana Hapsari, Dewi Indah Vebriyanti, Hardiansyah and all personnel of the Research Center for Quality System and Testing Technology, Indonesian Institute of Sciences that were involved in the research.

Using the sample data in Jakarta area, we also published a paper (see Sumaedi et al. 2012) that described the sub-study on explaining the behavioural intention of paratransit passengers using different perspectives, which is the rationale-based behavioural intention, in this book.

Reference

Sumaedi, S., Bakti, I. G. M. Y., & Yarmen, M. (2012). The empirical study of public transport passengers' behavioral intentions: The roles of service quality, perceived sacrifice, perceived value, and satisfaction (Case study: Paratransit passengers in Jakarta, Indonesia). *International Journal for Traffic and Transport Engineering*, *2*(1), 83–97. UDC: 656.121.072(594).

Contents

Abstract

Traffic congestion is one of the most common problems faced by all major cities in the world. It has triggered various interrelated effects, such as air and sound pollutions, climate change, health issues and fossil fuel scarcity. Surprisingly, even 'high evolving' technologies implemented in many countries had failed to cope with it. The literatures show that the failure is caused by the heavily skewed policy towards technical solutions while the human aspect of a consumer was simply overlooked.

Traffic congestion happens due to the unbalanced growth between the amount of the vehicles and road capacity. This phenomenon is the result of the passenger switching from public transport services to private vehicles. From the view of consumer behaviour discipline, this points out the lack of favourable behavioural intentions of the public transport passengers. Unfortunately, the literatures that discuss about public transport passengers' behavioural intentions—indicated by repetitive usage and WOM—are very limited. On the other hand, comprehending public transport passengers' behavioural intention is critical for policy makers in order to create friendly policies to hold down the growth of private vehicles. Without understanding the behavioural intentions of public transport passengers, government policies will end up to be near futile.

Evidently, public transport passengers' behavioural intentions need to be learned, assessed and discussed. There are two fundamental questions circling public transport passengers' behavioural intentions. First, what factors affect public transport passengers' behavioural intentions and second, which of those factors deemed to be the most contributing in forming public transport passengers' behavioural intentions. Therefore, this research aims to develop a model for explaining public transport passengers' behavioural intention.

This research proposed a model which explains the relationship between behavioural intention and its antecedents, i.e., customer satisfaction, perceived service quality, perceived sacrifice, trust, image and perceived value. This research was conducted in order to test the model using Indonesian paratransit passengers in five big cities (Jakarta, Bogor, Depok, Tangerang and Bekasi). Research design adopted was field survey using questionnaires. The respondents of the research are 563 paratransit passengers in Jabodetabek. Structural Equation Modelling was performed to analyse the data.

The result of data analysis shows that three factors affecting behavioural intentions are perceived sacrifice, image and perceived value of public transport passengers. In addition, this research also found that the most contributing factor in forming public transport passengers' behavioural intentions is image. Theoretical and practical implications are presented based on research findings. Furthermore, policies based on consumer behaviour are also formulated to aid the local government.

Keywords Consumer behaviour · Behavioural intention · Public transport · Paratransit · Loyalty

Chapter 1
Introduction

Abstract This chapter provides the research background to why this research was conducted, research questions and research purposes. In detail, the chapter describes the underlying causes of traffic congestion, the role of paratransit in Jabodetabek, traffic congestions from the perspective of consumer behaviour, research gaps related to passengers' behavioural intentions, research purposes and the organization of the book.

Keywords Traffic congestion · Paratransit · Consumer behaviour · Behavioural intention

1.1 Background

Global warming is a quite popular topic nowadays. The phrase 'global warming' serves its purpose when it comes to the rise of Earth's temperature due to excess amount of carbon dioxide (CO_2), methane (CH_4), nitrogen-oxide (NOx) (Ramanathan 1998; Kirby 2008). Global warming has unfriendly effects on Earth ecosystem; unstable climate, widespread melting of snow and ice and increased sea levels (Wang and Chameides 2005). Those effects are not the end, but the beginning of other devastating effects.

One of the causes of global warming is increase in number of carbon dioxide (CO_2) in atmosphere and one of the main contributors of the gas is transportation activities. Developed countries' transportation activities pour 3,386 million metric ton of CO_2 while developing countries contribute 1,374 million metric ton of CO_2 (The World Bank 2007)

Total emission of CO_2, which is poured by transportation activities, is contributed by road transportation emission (80 %), air transportation emission (13 %), water transportation emission (7 %) and train emission (0.5 %) (Kirby 2008). The data indicated that road transportation supplied the biggest part of the pie.

S. Sumaedi et al., *Public Transport Passengers' Behavioural Intentions*,
SpringerBriefs in Business, DOI: 10.1007/978-981-4585-24-8_1,

The amount of CO_2 resulted from transportation activities rises along the increasing numbers of vehicles on the streets (Kirby 2008). This condition becomes more severe every time there was traffic congestion. Congestion causes vehicle run in low speed in which it burns more fuels due to non-optimal works of the engine (Wijayanto 2009).

Besides global warming, traffic congestion also induces other problems, such as air and sound pollutions, health issues, fossil fuel scarcity and social problems (Cox 2010). Thus, traffic congestion needs to be dealt with.

Traffic congestion seems become an inevitable problem in major cities in the world, such as Tokyo, Los Angeles, Sao Paulo, Bangkok, Moscow, Shanghai, Mumbai, Mexico, New York and Seoul regardless their advanced transportation technology and systems (Tanaboriboon 1993; Cain and Kolpakov 2007; Pucher et al. 2005; Wang et al. 2012; Peng et al. 2012; UNEP 2010; Rivera et al. 2011; The Economist Intelligence Unit 2011; Downs 2004; PWC 2012). For example, Japan has monorails and the Shinkansen, but Tokyo is still overcrowded by vehicles (PWC 2012). New Delhi's Mass Rapid Transit and Bus Rapid Transit also appeared to be less effective than the predetermined expectations. Cox (2010) noted that Delhi Metro fell behind its target of 2 million pax per day; carrying only 250,000. The Bus Rapid Transit was not only failed to attract new customers, but also being labelled as 'road thief' (Cox 2010).

The technology-based policies fell short at part because it was not in tune with consumers' needs. Using the data of New Delhi's MRT and BRT, Cox (2010) reported that the underlying reason for the disappointment was the ill-suited pair of transportation system and customers' need for "mobility" and "transportation." Cox (2010) stated that the transportation system in New Delhi did not favour the needs of passengers with huge amount of luggage; people from informal sectors. As can be inferred from previous sentences, research on traffic congestion using consumer behaviour approach became important.

1.2 Traffic Congestion in Jabodetabek

Jabodetabek consists of DKI Jakarta, the capital city of Indonesia, and eight local governments (Kabupaten and Kota Bogor, Kota Depok, Kabupaten and Kota Tangerang, Kota Tangerang Selatan, and Kabupaten and Kota Bekasi. This 7.297 km^2 area has roughly 27.9 million inhabitants (BPS 2011). Jabodetabek is currently the most important and strategic area in Indonesia because of its economic activities with Rp 351,000 billion Gross Regional Domestic Product which was simply 22 % of Indonesia's Gross Domestic Product in 2002 (JICA Study Team 2004).

Related to transportation, Jabodetabek has a problem like any other major cities in the world, traffic congestion. Traffic congestion happens in peak hours at morning and evening. The congestion has inflicted Rp 5.5 billion of financial lost in form of vehicle operational cost and travel time (JICA Study Team 2004).

The congestion in Jabodetabek is caused by the increase in traffic activities in the area (JICA Study Team 2004). Besides that, the congestion is also caused by the decline in public transport usage and the increase usage of private vehicles. The research carried out by JUTPI (in Prakesit 2011) shows that public transport usage declined from 38.8 % in 2002 to 12.9 % in 2010. Conversely, according to the Traffic Directorate of Polda Metro Jaya, the usage of private vehicles increases year to year. The amount of private vehicles has reached 6.3 million units or equal to 98 % of the total amount of vehicles in Jabodetabek (UTIC 2010). Ironically, private vehicles can only satisfy 44 % of Jakarta's total mobility needs while the mere amounts can fulfil 56 % of Jakarta's mobility needs (DTKJ [Dewan Transportasi Kota Jakarta] 2012).

Currently, local government in Jabodetabek is focusing on how to attract people to use public transport. The local government is trying to improve the performance of their public transport by introducing new mode of public transportation which is safer, faster and more comfortable. For example, the local government of DKI Jakarta is developing and implementing Bus Rapid Transit (BRT) called Trans-Jakarta. They are also planning other kinds of public transportation system like Monorail and Mass Rapid Transit (MRT). Local government expects to attract public interest by improving the performance of public transport so that the amount of vehicle circulated in the road can be declined. Furthermore, it is expected that the congestion problem will be solved by the increase in of the use of public transport. Hence, it can be concluded that the problem of traffic congestion in Jabodetabek is closely related to the usage of public transport.

1.3 Paratransit in Jabodetabek, Indonesia

Public transport in Jabodetabek is dominated by paratransit. The amount of paratransit in Jabodetabek can be seen in Table 1.1. According to the Decree of Minister of Transportation No. 35/2003, paratransit is a mean of transport from one place to another inside a city or district or inside the capital city of Jakarta using public bus or public passenger car specified for certain routes. Public bus includes big-size bus, medium-size bus and small-size bus. The decree also stated that for a city with more than 500,000 inhabitants like Jabodetabek, big-size bus provide service in main routes, medium-size bus provide service for branch routes and small-size bus or public passenger car provide service for twig routes.

Paratransit is one of the most expected public transports to help the mobility of people inside the city. The superior feature of the paratransit is it has a various route, from twig routes to main routes, so that the paratransit is expected to cover wide range of people mobility needs. Furthermore, the amount of paratransit in Jabodetabek increases from year to year (Directorate General of Land Transportation 2011). This makes people to access the service easier.

Currently, the local government is developing new transportation system like Mass Rapid Transit and Bus Rapid Transit. However, it does not mean that

Table 1.1 The amount of paratransit in Jabodetabek in 2010

City	The amount of paratransit in 2010 (Unit)			
	Big-size bus	Medium-size bus	Small-sized bus	Public passenger car
DKI Jakarta	2.809	7.821	26.002	2.576
Bogor	239	73	4.529	3.459
Depok	157	46	568	2.880
Tangerang	89	–	300	2.468
Bekasi	264	211	–	1.732

Directorate General of Land Transportation 2011

paratransit is no longer needed. Its wide coverage is still cannot be matched by MRT or BRT since it can sink deep into small streets all over the city. Thus, it can be integrated with both MRT and BRT, whereas paratransit can act as feeder for both MRT and BRT. Hence, the existence of sufficient paratransit is still needed. Therefore, research on public transport usage should focus on paratransit usage.

1.4 Traffic Congestion from the Perspective of Consumer Behaviour

Congestion refers to a condition when the traffic is stagnated or stalled because the number of vehicles exceeds road capacity (Litman 2011). Like what we have discussed in the previous section (Sect. 1.2), congestion in Jabodetabek happens because people prefer using private vehicle than utilizing public transport. In other word, traffic congestion is caused by people's usage pattern on public transport.

In the perspective of consumer behaviour, the low usage of public transport can be explained by the concept of behavioural intention. This concept is also called as loyalty intention (Olsen et al. 2013; Chiou et al. 2009; Chiou and Shen 2006; Yen and Lu 2008) or attitudinal loyalty (Sanzo et al. 2007; Jaiswal and Niraj 2011; Aurier and Lanauze 2011, 2012). Behavioural intention illustrates the extent of repurchase intention and the willingness to engage in word of mouth communication (Bitner 1990; Nadiri et al. 2008; Saha and Theingi 2008). Favourable behavioural intention will lead to higher repurchase intention and elevated inclination to recommend the product to others and vice versa. Furthermore, behavioural intention will also create positive impact on product's market share and sales (Clemes et al. 2008).

In the context of public transport, passenger with favourable behavioural intention will keep using public transport and even recommend it to others (Lai and Chen 2011) whereas passenger with unfavourable behavioural intention will tend to switch to other affordable alternatives. In Jabodetabek, people with unfavourable behavioural intention have quite wide range of alternatives like motorcycle and private car since both can be obtained easily using affordable credit instalment. This supportive financial arrangement makes people with unfavourable behavioural intention chose to buy private cars and motorcycles.

Based on above explanation, we can conclude that traffic congestion is caused by the failure of public transport in creating favourable behavioural intention. People with unfavourable behaviour intention tend to switch to other "means of transportation" and very unlikely to become "informal marketing" agent for public transport. Local government in Jabodetabek should strive to increase public transport passengers' behavioural intention. Hence, the knowledge on the antecedent factors of passengers' behavioural intention and the most contributing factors in forming behavioural intention become important and necessary.

1.5 Research Gaps

Studies on passengers' behavioural intentions of public land transport are fairly limited and mostly carried out in the context of Taiwan (Sumaedi et al. 2012). For example, Lai and Chen (2011) studied the behavioural intentions of Kaohsiung Mass Rapid Transit (KMRT) passenger in Taiwan. They investigated the relationship of behavioural intentions with few latent factors, such as public transit involvement, service quality, perceived value and satisfaction. Their study resulted on a model of behavioural intentions that explains the relationship between public transit involvement, service quality, perceived value, satisfaction and behavioural intentions.

The research conducted by Wen et al. (2005) was intended to identify factors that affect the loyalty of intercity bus passengers in Taiwan. Wen et al. (2005) has developed a model containing the relationship between eight variables (satisfaction, service value, service quality, sacrifice, switching cost, attractiveness of competitors, and trust) and the loyalty of the bus passengers.

Different with two previous researches, Jen and Hu (2003) only learned about passengers' repurchase intentions in Taiwan. In their study, Jen and Hu (2003) built and tested perceived value model in order to identify factors that affect repurchase intentions of bus passengers. Jen and Hu (2003) combined also other variables beside perceived value and repurchase intentions, such as service quality, perceived benefits, perceived non-monetary price, perceived monetary price, perceived costs and attractiveness of alternative modes.

As discussed before, the previous researches on passengers' behavioural intentions already developed models that can explain the behavioural intentions of public transport passengers. However, the fact that their studies were conducted in Taiwan makes the results dubious for Indonesian context since the two have different cultures. It is generally accepted that researches on management science can generate various results and theoretical implications because of differences in culture or other contextual factors (Nair 2006; Bhaskaran and Sukumaran 2007).

Besides the previous explanation, few gaps also found in previous researches. First, previous researches were dominated by the paradigm that see that behavioural intention is the result of rational decision making and considerations of all available alternatives (Jen and Hu 2003; Wen et al. 2005). Thus, they believed that customers are not only evaluating public transport, but also collecting as many information as possible about all available alternatives that can be used as comparison with their evaluation result on public transport (Solomon 2012). This assumption has made previous researches used variables related to the evaluation results of available alternatives/competitors, such as, attractiveness of alternative modes/competitor and switching cost (Jen and Hu 2003; Wen et al. 2005).

The assumption is mostly debatable since literatures on consumer behaviour show that behavioural intention is a result of irrational decision-making process whereas the decision to be loyal to a product shows inertia after first purchase or emotional bonding to a product purchased (Solomon 2012). Favourable behavioural intentions tend to make most customers ignore the existence of available alternatives/competitor. Gummesson (2008) stated that consumer's loyalty make consumers' perception on the existence of other alternatives decreases and not the other way around. Furthermore, literatures show that a loyal consumer, brand zealot, treats the brand as part of himself or herself (Aggarwal 2004). He or she will not switch to other rational alternatives. The relationship between a consumer with favourable behavioural intention and his or her chosen products is similar with the relationship between man and wife. It makes them insensitive toward other men or women (Solomon 2012). Without any intention to switch, consumer is very unlikely to search other alternatives products available at the market. Thus, variables related to the evaluation of alternative products or competitor become irrelevant in discussing the growth of behavioural intention.

The paper of Earl (2012) shows that consumer makes decision regarding his consumption of 'automotive product,' which serves similar functional utilitarian with public transport services, tend to be irrational and do not really compare the costs and benefits gained by using other means of transport. Furthermore, he also found that consumer chose to buy vehicle rather that rent one although renting is more beneficial than buying. Thus, passengers' behavioural intentions should be seen as irrational decision making processes based on emotional bound between consumers and their chosen products, rather than a rational decision-making processes involving thorough evaluations of their chosen products as well as other alternatives (Solomon 2012).

Second, the conceptual models developed in previous researches see the behavioural intentions as the results of consumers' cognitive assessment (Lai and Chen 2011; Wen et al. 2005; Jen and Hu 2003). This point of view has made the previous conceptual models dominated by variable representing the performance of service provider in fulfilling functional utilitarian needs, which is service quality, as the initial cause of behavioural intention (Lai and Chen 2011; Wen et al. 2005; Jen and Hu 2003). Thus, the consequence is that passengers' behavioural intention is believed as a function that always involves service quality. Furthermore, the improvement of service quality is expected to have linear and positive effect on behavioural intentions directly or indirectly through psychological variables, such as, satisfaction (Lai and Chen 2011; Wen et al. 2005; Jen and Hu

2003), value (Lai and Chen 2011; Wen et al. 2005; Jen and Hu 2003) and trust (Wen et al. 2005).

Based on the conceptual model, service quality always becomes a key variable in increasing public transport passengers' behavioural intentions. The practitioners of public transport are then recommended to improve service quality in order to achieve better passengers' behavioural intention. However, the efforts to improve service quality like in the case of TransJakarta and Metro Delhi often fell short.

Literatures on consumer behaviour show that human needs stem not only from functional utilitarian needs that is cognitively evaluated, but also from hedonic experiential need which is affectively evaluated (Aggarwal 2004; Falk et al. 2010; Solomon 2012). The functional utilitarian needs are related to instrumental and functional aspects of a product (i.e. service quality), while experiential hedonic needs emphasize on emotional aspect of the product (Aggarwal 2004; Falk et al. 2010; Solomon 2012). It is well known the notion 'you are what you consume', which is illustrates how a product functions as a fulfilment of hedonic experiential needs (Solomon 2012).

Public transports fulfil similar functional utilitarian needs with private vehicle, such as car or motorcycle that produce by automotive company. Further, the public transports compete with private vehicles in attracting consumers' interests indeed. Shimp (2010) shows that there are often automotive ads that did not expose vehicle's ability to fulfil functional utilitarian needs but only show vehicles advantages in satisfying hedonic experiential needs. Researches by Zhou et al. (2012), Keaveney et al. (2012), Earl (2012) also confirmed the role of hedonic variable toward automotive consumer loyalty.

Based on previous description, it is important to include hedonic variable into passengers' behavioural intentions model. In other words, researches on passengers' behavioural intention need to view that there is not only the variables that represent the performance of public transport on satisfying functional utilitarian needs and evaluated cognitively (service quality), but also the variables that represent the performance of public transport on satisfying hedonic experiential needs and evaluated affectively has role to initiate the formation of favourable passengers' behavioural intention.

Third, previous researches disregarded the fact that the result of cognitive evaluation, i.e. service quality, can be affected by product information obtained by the consumers before they consumed the consumed. The consumers tend to evaluate the information as a part of the product they consumed, whereas positive information would very likely lead to positive result of cognitive evaluation and vice versa (Wilcox et al. 2011). In other words, consumers can provide different result of cognitive evaluation for a same product depends on the information they obtained before. Regarding to this topic, an experimental research that was often carried out is the research that shows how consumers evaluate cola product differently when they know and don't know the brand of the product before (Solomon 2012). This condition can also happen in the case of public transport passengers' cognitive evaluation in Indonesia considering the phenomenon shows that automotive product information, which is the image of a brand, is used as a quality indicator. For example, Xenia (Daihatsu) and Avanza (Toyota) are very similar to each other and in fact, they are the result of collaboration between Daihatsu and Toyota. However, people still perceived Avanza as a better product because Toyota has superior brand image than Daihatsu.

The three research gaps mentioned above can be solved by including image as one of the variables in the behavioural intentions conceptual model. Image is a variable that represent the fulfilment of experiential hedonic needs and it is evaluated using affective (emotional) evaluation process (Falk et al. 2010). Image is often used by consumer in decision-making process (Solomon 2012) and it can affect the results of cognitive evaluation, service quality in this case, if it has been known before a product is consumed (Wilcox et al. 2011). In the context of public land transport, Wen et al. (2005) suggested this variable to be included in the future behavioural intention model. However, until now, there are no researchers that have done so. Therefore, it is important to include image as one of the variables in passengers' behavioural intentions model.

1.6 Research Questions

Based on previous sections, generally, this research tries to answer one fundamental question, "what kind of behavioural intentions model that can explain paratransit passengers' behavioural intentions in Jabodetabek, Indonesia?"

Previous researches already identified some influencing variables, such as, satisfaction, perceived service quality, perceived sacrifice, trust image and perceived value. Based on that consideration, this research also covers those variables. Furthermore, based on research gaps found, the role of image was also tested. Thus, more specifically, this research was conducted to answer the question "how is the relationship between consumer satisfactions, perceived service quality, perceived sacrifice, trust, image, perceived value and behavioural intentions of Jabodetabek paratransit passengers?".

1.7 Research Purposes

Generally, the purpose of this research is to develop conceptual model that can explain behavioural intentions of paratransit passengers in Jabodetabek, Indonesia.

In details, the purposes are to examine:

- The influence of perceived sacrifice on perceived value of paratransit passengers in Jabodetabek
- The influence of image on service quality, satisfaction, perceived value, behavioural intentions and trust of paratransit passengers in Jabodetabek

- The influence of perceived service quality on satisfaction, perceived value and behavioural intentions of paratransit passengers in Jabodetabek
- The influence of perceived value on satisfaction, trust and behavioural intentions of paratransit passengers in Jabodetabek
- The influence of satisfaction on trust and behavioural intentions of paratransit passengers in Jabodetabek
- The influence of trust on behavioural intentions of paratransit passengers in Jabodetabek.

1.8 Organisation

This book consists of seven chapters. The explanation is as follows. This chapter provides research background that explains the reasons of why this research was conducted, research questions and research purposes. In detail, this chapter describes the underlying causes of traffic congestion, the role of paratransit in Jabodetabek, traffic congestions from the perspective of consumer behaviour, research gaps related to passengers' behavioural intentions, research purposes and the organisation of the book.

Chapter 2 discusses theoretical background related to this research. This chapter consists of explanation on the concept of behavioural intention, behavioural intention as a decision-making process and antecedents of behavioural intention. Aligned with research purposes, the antecedents of behavioural intention include satisfaction, service quality, perceived sacrifice, trust, image and perceived value.

Chapter 3 presents conceptual model and hypotheses which will be tested later. Chapter 3 also explains the arguments used for building the model and its hypotheses from both theoretical and empirical sides.

Chapter 4 explicates research method used in this study. It contains research object (paratransit in Jabodetabek), research population and sample, operational definition and scale of variables, pilot test, survey and data analysis method.

Research results are described in Chap. 5. It comprises the result of confirmatory factor analysis and the upshot of structural model test.

Chapter 6 relates the implication of this research. Implications conferred in Chap. 6 are theoretical, managerial, and policy implications. Finally, Chap. 7 presents the conclusions and limitations of the research and next research agenda.

References

Aggarwal, P. (2004). The effects of brand relationship norms on consumer attitudes and behavior. *Journal of Consumer Research, 31*, 87–101. Retrieved from http://www.utsc.utoronto.ca/~aggarwal/BrandRelationshipNorms.pdf

Aurier, P., & Lanauze, G. S. D. (2011). Impacts of in-store manufacturer brand expression on perceived value, relationship quality and attitudinal loyalty. *International Journal of Retail and Distribution Management, 39*(11), 810–835.

Aurier, P., & Lanauze, G. S. D. (2012). Impacts of perceived brand relationship orientation on attitudinal loyalty: An application to strong brands in the packaged goods sector. *European Journal of Marketing, 46*(11), 1602–1627.

Bhaskaran, S., & Sukumaran, N. (2007). National culture, business culture and management practices: Consequential relationships? *Cross Cultural Management: An International Journal, 14*(1), 54–67.

Bitner, M. J. (1990). Evaluating service encounters: The effect of physical surroundings and employee responses. *Journal of Marketing, 54*, 69–82.

BPS. (2011). *Development number of motor vehicle by type in 1987–2011*. Retrieved from February 6, 2013 (GMT + 7 14:00:00), from http://www.bps.go.id/tab_sub/view.php?tabel=1&daftar=1&id_subyek=17&notab=12

Cain, A., & Kolpakov, A. (2007). Developing effective mass transit systems. *Proceedings of the 5th International Workshop on Public Transportation*. Center for Urban Transportation Research, University of South Florida.

Chiou, J. S., & Shen, C. C. (2006). The effects of satisfaction, opportunism, and asset specificity on consumers' loyalty intention toward internet portal sites. *International Journal of Service Industry Management, 17*(1), 7–22.

Chiou, J. S., Wu, L. T., & Sung, Y. P. (2009). Buyer satisfaction and loyalty intention in online auctions: Online auction web site versus online auction seller. *Journal of Service Management, 20*(5), 521–543.

Clemes, M. D., Gan, C., Kao, T. H., & Choong, M. (2008). An empirical analysis of customer satisfaction in international air travel. *Innovative Marketing, 4*, 50–62.

Cox, P. (2010). *Moving people, sustainable transport development*. Cape Town: UCT Press.

Directorate General of Land Transportation. (2011). *Perhubungan darat dalam angka 2010*. Jakarta: Directorate General of Land Transportation.

Downs, A. (2004). Why traffic congestion is here to stay… and will get worse. *Access, 25*, 19–25.

DTKJ (Dewan Transportasi Kota Jakarta). (2012). *Ibukota yang tak punya rencana*. Retrieved from http://dtk-jakarta.or.id/v3/en/slide-view/ibukota-negara-yang-tak-punya-rencana/

Earl, P. E. (2012). Experiential analysis of automotive consumption. *Journal of Business Research, 65*, 1067–1072.

Falk, T., Hammerschmidt, M., & Schepers, J. J. L. (2010). The service quality satisfaction link revisited: exploring a symmetries and dynamics. *Journal of the Academy of Marketing Science, 38*(3), 288–302.

Gummesson, E. (2008). *Total relationship marketing* (3rd ed.). Oxford: Butterworth-Heinemann.

Jaiswal, A. K., & Niraj, R. (2011). Examining mediating role of attitudinal loyalty and nonlinear effects in satisfaction-behavioral intentions relationship. *Journal of Services Marketing, 25*(3), 165–175.

Jen, W., & Hu, K. C. (2003). Application of perceived value model to identify factors affecting passengers' repurchases intention on city bus: A case of the Taipei metropolitan area. *Transportation, 30*, 307–327.

JICA StudyTeam. (2004). *The Study on Integrated Transportation Master Plan for Jabodetabek (Phase II), Final Report, Main Report, 1: Master Plan Study, in coordination with National Development Planning Agency (BAPPENAS) Republic Indonesia.*

Keaveney, S. M., Hernman, A., Befurt, R., & Landwehr, J. R. (2012). The eyes have it: How a car's face influences consumer categorization and evaluation of product line extensions. *Psychology and Marketing, 29*(1), 36–51.

Kirby, A. (2008). *CCCC kick the habit: A UN guide to climate neutrality*. Nairobi: UNEP.

Lai, W.-T., & Chen, C.-F. (2011). Behavioral intention of public transit passenger—the role of service quality, perceived value, satisfaction and involvement. *Transport Policy, 18*, 318–325.

Litman, T. (2011). Generated traffic and induced travel implications for transport planning. *ITE Journal, 71*(4), 38–47.

Nadiri, H., Hussain, K., Ekiz, E. H., & Erdoğan, S. (2008). An investigation on the factors influencing passengers' loyalty in the North Cyprus national airline. *The TQM Journal, 20*(3), 265–280.

Nair, A. (2006). Meta-analysis of the relationship between quality management practices and firm performance-implications for quality management theory development. *Journal of Operations Management, 24*(6), 948–975.

Olsen, S. O., Tudoran, A. A., Brunso, K., & Verbeke, W. (2013). Extending the prevalent consumer loyalty modelling: The role of habit strength. *European Journal of Marketing, 47*(1/2), 303–323.

Peng, Z. R., Sun, J.,& Lu, Q. C. (2012). China's public transportation: Problems, policies, and prospective of sustainability. *ITE Journal, 82*(5), 36–60.

Prakesit, D. (2011). Menyelamatkan Jakarta dari kemacetan total. *Jurnal Prakasa Infrastruktur Indonesia, 6*, 8–13.

Pucher, J., Park, H., & Kim, M. H. (2005). Public transport in Seoul: Meeting the burgeoning travel demands of a megacity. *Public Transport International, 3*, 54–60.

PWC. (2012). *Cities of opportunity*. New York: The Partnership for New York City.

Ramanathan, V. (1998). Trace-gas greenhouse effect and global warming. *AMBIO, 27*(3), 187–197.

Rivera, V. I. G., Salvador, H., Osorno, J. A. A., Zaragoza, M. L., & Ruvalcaba, J. I. (2011). Implementing sustainable urban travel policies in Mexico, Discussion paper 14, *International Transport Forum.*

Saha, G. C., & Theingi. (2008). Service quality, satisfaction, and behavioral intentions: A study of low-cost airline carriers in Thailand. *Managing Service Quality, 19* (3), 350–372.

Sanzo, M. J., Santos, M. L., Alvarez, L. I., & Vazquez, R. (2007). The effect of a buyer's market orientation on attitudinal loyalty toward a supplier: Is dependence a moderator? *Supply Chain Management, 12*(4), 267–283.

Shimp, T. A. (2010). *Integrated marketing communication in advertising and promotion* (8th ed.). Beijing: South-Western Cengage Learning.

Solomon, M. R. (2012). *Consumer behavior; Buying, having and being* (10th ed.). London: Prentice Hall.

Sumaedi, S., Bakti, I. G. M. Y. & Yarmen, M. (2012). The empirical study of public transport passengers' behavioural intentions: The roles of service quality, perceived sacrifice, perceived value, and satisfaction (case study: Paratransit passengers in Jakarta, Indonesia). *Interational Journal for Traffic and Transport Engineering*, UDC: 656.121.072(594). 2(1), 83–97.

Tanaboriboon, Y. (1993). Bangkok traffic. *IATSS Research, 17*(1), 14–23.

The Economist Intelligence Unit. (2011). *Making up for lost time: Public transportation in Brazil's metropolitan areas: A report from the Economist Intelligence Unit*. Retrieved from http://www.managementthinking.eiu.com/sites/default/files/downloads/Public%20transport%20in%20Brazil_0.pdf

The Word Bank. (2007). The Little Green Data Book 2007. The International Bank for Reconstruction and Development, Washington, USA.

UNEP. (2010). *UNEP Environmental Assessment : EXPO 2010*, Shanghai, China. Retrieved from http://www.unep.org/pdf/SHANGHAI_REPORT_FullReport.pdf

UTIC. (2010). *Transportasi Kota Jakarta mengkhawatirkan.* Retrieved from http://bstp.hubdat.web.id/index.php?mod=detilSorotan&idMenuKiri=345&idSorotan=54

Wang, M., Martin, E., & Shaheen, S. (2012). Carsharing in Shanghai, China: Analysis of Behavioral Response to a Local Survey and Potential Competition. *TRB Annual Meeting*. Retrieved from http://tsrc.berkeley.edu/carsharingshanghai

Wang, J., & Chameides, B. (2005). Global Warmings's Increasingly Visible Impact, Environmental Defense. Los Angeles. Retrieved from http://www.edf.org/sites/default/files/4891_GlobalWarmingImpacts.pdf

Wen, C.-H., Lan, L. W., & Cheng, H.-L. (2005). Structural equation modelling to determine passenger loyalty toward intercity bus services. *Journal of the Transportation Research Board, 1927*, 249–255.

Wijayanto, Y. (2009). Analysis of vehicle speed on Brigjen Sudiarto (Majapahit) street, Semarang City and the effect on fuel oil consumption. Thesis, Faculty of Engineering, Diponegoro University [in Indonesian]. Retrieved from http://digilib.usm.ac.id/files/disk1/5/gdl-usm–mudjiastut-201-1-analisa-n.pdf

Wilcox, K., Roggenveen, A. L., & Grewal, D. (2011). Shall I tell you now or later? Assimilation and contrast in the evaluation of experiential products. *Journal of Consumer Research, 38*(4), 763–773.

Yen, C. H., & Lu, P. H. (2008). Effects of e-service quality on loyalty intention: An empirical study in online auction. *Managing Service Quality, 18*(2), 127–146.

Zhou, Z., Zhang, Q., Su, C., & Zhou, N. (2012). How do brand communities generate brand relationships? Intermediate mechanisms. *Journal of Business Research, 65*, 890–895.

Chapter 2
Literature Review

Abstract This chapter discusses the theoretical background related to this research. It consists of an explanation on the concept of behavioural intention, behavioural intention as a decision making process and antecedents of behavioural intention. Aligned with the research purposes, the antecedents of behavioural intention include satisfaction, service quality, perceived sacrifice, trust, image and perceived value.

Keywords Behavioural intention · Satisfaction · Service quality · Perceived sacrifice · Trust · Image · Perceive value

2.1 Introduction

Before 1930s, companies did not think much of consumer behaviour and efforts to understand it had not been encouraged. Firms' sole purpose was to produce (Lynd 1936). Adam Smith avowed this statement in his explanation on mercantilism which put the needs of the producers above all else (Kucukaksoy 2011).

After Georgescu-Roegen (1936) analyzed *the theory of choice*, entrepreneurs started to think of how to understand consumer behaviour and use it as marketing weapon (Gowdy and Mesner 1998). Georgescu-Roegen (1936) stated that transactions happened in the market is not as simple as supply and demand curves which only consider price and quantity. In the same year, Lynd (1936) also showed how supply and demand curves are irrelevant because of their underlying assumptions. Furthermore, Lynd (1936) argued that productions should be directed by consumer behaviour because goods scarcity is no longer an issue since the spread of production machinery, confusing consumer rationality with dynamic needs, and the fact that the consumers seemed to enjoy their instability.

'*Consumer behaviour refers to the behaviour that consumers display in searching for, purchasing, using, evaluating, and disposing of products and services that they expect will satisfy their needs* (Schiffman and Kanuk 1994)'. Engel et al. (1994)

S. Sumaedi et al., *Public Transport Passengers' Behavioural Intentions*, SpringerBriefs in Business, DOI: 10.1007/978-981-4585-24-8_2,

defined consumer behaviour as '*behaviours that are directly involved in the action of obtaining, consuming, and spending our products and services, including the decision processes that precede and follow these actions*'. Meanwhile Solomon (2012) stated that '*the field of consumer behaviour covers a lot of ground: It is the study of the processes involved when individuals or groups select, purchase, use, or dispose of products, services, ideas, or experiences to satisfy needs and desires*'.

One of the study subjects of consumer behaviour is consumer behavioural intention. Nowadays, it becomes one of the most widely discussed subjects in marketing literature. This is not surprising since the ability to well-managed consumer behavioural intention creates competitive advantage for firms. Literature shows that favourable behavioural intention results on insensitivity towards price (Lai and Chen 2011) and attractiveness of competitor (Gummeson 2008). Furthermore, behavioural intention is believed to be a mediating variable between customer satisfaction and economic benefits like profitability, market share and income (Clemes et al. 2008). Firms that can satisfy their customers but fail to create favourable behavioural intention among their customers will not obtain significant profits.

2.2 Understanding Behavioural Intention

According to the theory of planned behaviour (TPB), behavioural intention is a predictor of future behaviour (Ajzen 1991). Zeithaml et al. (1996) divide behavioural intention into five, which are: loyalty, switch, pay more, external responses and internal responses. Bush et al. (2004) argued that behavioural intention consists of brand loyalty, positive word of mouth (WOM), and switching and complaining behaviour. Furthermore, Bloemer et al. (1999) broke up behavioural intention to WOM, purchase intention, price sensitivity and complaint behaviour. Yang and Peterson (2004) concluded that most researchers view behavioural intention represents customer loyalty. Fornell (1992) explained that customer loyalty is formed by two kinds of intentions; repurchase intentions and the intention to recommend the product to others. Thus, the concept of behavioural intention proposed by Yang and Peterson (2004) and Fornell (1992) refers to the intention to repurchase and recommend. Further, the concept of behavioural intention is also referred as attitudinal loyalty (Sanzo et al. 2007; Jaiswal and Niraj 2011; Aurier and Lanauze 2011, 2012) or loyalty intention (Olsen et al. 2013; Chiou et al. 2009; Chiou and Shen 2006; Yen and Lu 2008).

In the context of public land transport services, some researchers already conducted study on passengers' behavioural intention, such as Jen and Hu (2003), Wen et al. (2005), and Lai and Chen (2011). They unanimously agree that behavioural intention represents customer loyalty, which consists of the intention to repurchase a product/service and recommend it to others. Thus, passengers' behavioural intention represents to what extent public transportation passengers will reuse and recommend public transport (Jen and Hu 2003; Lai and Chen 2011). The more favourable behavioural intention of the passengers, the more their intention to reuse public transport and recommend it to others and vice versa.

2.3 Behavioural Intention as Consumer Decision Making Process

Behavioural intention reflects consumer decision-making process for keeping loyal to a product he or she consumed and even become 'informal' marketing agent for the product (Egan 2011; Solomon 2012). Generally, there are two paradigms concerning consumer decision-making process. First paradigm is the rational paradigm. The paradigm sees consumer decision-making process as a rational process which uses all available information and the decision is made based on 'cost-benefit' consideration for each alternative of decisions (Solomon 2012). Rational decision-making process involves some steps that are shown in Fig. 2.1.

Second paradigm is the irrational paradigm. The irrational paradigm explains why consumers often do not choose the rationally best product available at the market. The irrational paradigm consists of two perspectives, which are behavioural influence perspective and experiential perspective. The first perspective refers to the view that see decision-making process based on 'only' a learned response to environment cue while the second perspective believes that decision-making process is a subjective process and it is naturally hard to be explained, whereas consumers' decision is made mostly based on the emotional bond (Solomon 2012).

In accordance with previous Sect. 2.2 explanation, behavioural intention represents consumer loyalty towards certain product. The consequence of the concept is that consumers never feel like being forced to have relationship with the products they consumed. Consumers have favourable behavioural intention is not caused by lack of alternative. 'Lack of alternative' will keep customer to reuse the product even though he/she doesn't willing to recommend and even reuse it. More clearly, this research differentiates consumers with favourable behavioural intention and the ones with lack of alternative even though both might seem loyal to reuse certain products.

Based on the TPB proposed by Ajzen (1991), we argue that consumers who reuse product due to lack of alternative will certainly switch to other products if they have the ability to do so since the lost of their perceived behavioural control. On the other hand, consumers with favourable behavioural intention will perceive the existence of other alternatives as scarce and become insensitive with the cost-benefit considerations of other alternatives (Gummesson 2008). In other words, even though there are competitors those offers rationally better products, consumers with favourable behavioural intention will not change their preferences.

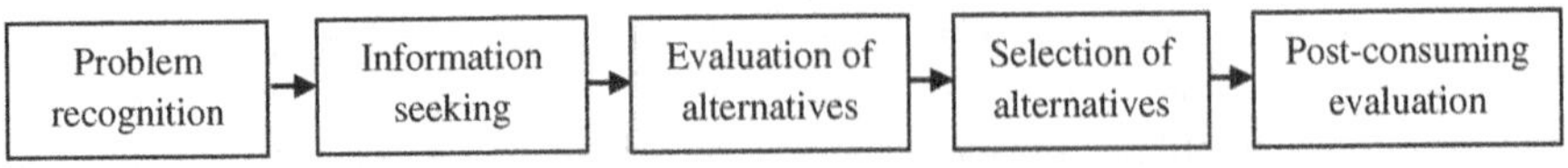

Fig. 2.1 Rational decision-making process (Mowen and dan Minor 2002; Solomon 2012)

Based on previous explanation, we can conclude that behavioural intention is an irrational decision-making process. Earl's (2012) auto ethnography research confirmed that consumers tend to make decision related to their consumption on automotive products, which serves the same functional utilitarian needs with public transports, irrationally and did not compare objective costs and benefits between alternatives. Furthermore, Earl (2012) found that consumers chose to buy vehicle rather than rent vehicle even though renting is rationally more beneficial than buying vehicle in the research context.

Consumer decision to be loyal to certain product is part of irrational decision-making process. However, whether it is due to inertia, 'we buy a brand out of habit merely because it requires less effort' (Solomon 2012), or emotional bond is still being debated among experts (Solomon, 2012). In relationship marketing literature, the view that favourable behavioural intention shows consumer inertia was popularized by Sheth and Parvatiyar (1995). However, Bagozzi (1995) and Peterson (1995) seemed to reject the argument. Furthermore, recent literature in relational marketing, a field that discusses how marketers develop relationship with customers, shows that favourable behavioural intention reflects emotional bond between consumers and their chosen products (Zhou et al. 2012). Thus, this research perceives public transport passengers' behavioural intention as the result of irrational decision-making process that suggests emotional bound between passengers and their chosen mode of transportation. Favourable behavioural intention indicates strong emotional bond between passengers and their chosen public transport. More favourable behavioural intention of the passengers shows stronger emotional bond between them and their chosen public transport and vice versa.

2.4 Antecedents of Behavioural Intention

Relationship marketing literature adopts the principles of success chain management to explain behavioural intention (Bruhn 2003). It stated that a consequence variable is the result of interrelated relationship of antecedent variables, just like a chain (Bruhn 2003). More clearly, consequence variable is affected by an antecedent variable which is also influenced by another antecedent variable and so on until we find the tip of all variables which supposed to represent valuation of service provider's input (Bruhn 2003). This research refers the tip of antecedent variables as 'initiation variable'. Thus, antecedents of behavioural intention consists of initiation variable and other antecedents variables.

In the context of behavioural intention, service provider's inputs are the responds of service provider on consumer apparent needs. At first, it was intended to fulfil only functional utilitarian needs which relate to products' instrumental and functional aspects (Falk et al. 2010). More clearly, service providers promised products' attributes and functional benefits, which are evaluated by consumers using cognitive evaluation, to meet the needs (Shimp 2010; Solomon, 2012). Related to this matter, marketing scholars, especially researchers on service marketing, then argued that service quality represents the performance of service providers in fulfilling functional utilitarian needs which were mostly evaluated using cognitive assessment rather than affective assessment (Chang and Wang, 2011; Dodds 2002). Therefore, it was not surprising when we found literatures on behavioural intention generally considered service quality as initiation variable (Lien and Yu 2001; Al-Rousan et al. 2010; Cronin et al. 2000; Liu et al. 2005; Huang, 2009) and it also happened in the context of public land transport (Lai and Chen 2011; Wen et al. 2005).

Literature studies on consumer behaviour have already identified that human needs develop beyond functional utilitarian needs to hedonic experiential needs (Solomon 2012). Hedonic experiential needs are related to the emotional aspects of product, such as how the product can become the part of consumers' life or how the product can show consumers' self-identity (Shimp 2010; Falk et al. 2010, Solomon 2012). Falk et al. (2010)'s research on quality-satisfaction relationship included variable that represents the performance of service provider in carrying out activities that fulfil consumers' hedonic needs aside from service quality. Integrated marketing communication literature even developed a marketing communication model called as Hedonic Experiential Model (HEM), which emphasizes the importance of integrated marketing communication that focus on product's emotional aspect (Shimp 2010). Thus, the input of service provider is no longer focused on the effort to satisfy functional utilitarian needs, but also hedonic experiential needs.

Image is a variable that represents the performance of service provider in satisfying hedonic experiential needs (Falk et al. 2010). Image illustrates what is in consumers' minds related to certain product/brand/object/service provider (Sondoh Jr et al. 2007; Andreassen and Lindestad 1998; Bloemer and de Ruyter 1998; Bridson and Mavando 2011). Unlike service quality, image is evaluated by consumers using affective assessment (Zins 20010; Andreassen and Lindestad 1998).

To our knowledge, research on public land transport passengers' behavioural intention that involves 'image' as one of the in the context of land transportation was nonexistent (Wen et al. 2005). Even so, literature on consumer behaviour shows that product image can form emotional bound which is the root of favourable behavioural intention (Kandampully and Hu 2007; Nguyen and Leblanc 1998, Chen and Tsai 2007; Ryu et al. 2008). Furthermore, in the context of automotive industry, which serves the same functional utilitarian needs with public transports, Shimp (2010) cited the existence of ads that are more dominated by efforts to accentuate the image of the products than their functional features. Relational marketing literature also denotes how Harley Davidson emotionally bounded to the brand image (Pugliese and Cagan 2002). Empirically, researchers also confirmed the important role of image in the formation of emotional bound with automotive products (Zhou et al. 2012). Therefore, it was understandable when Wen et al. (2005) suggested 'image' as antecedent variable of public land transport passengers' behavioural intention.

It is generally known in the practice of marketing that service provider not only responded consumers' needs, but also charged monetary price and even forced consumers to make non-monetary sacrifices, such as time (Zeithaml 1988). Marketers have long used the formula of 4P marketing mix (product, place, promotion and price) (Gronroos 1994). The management of price and place is an attempt to make 'sacrifices' made by consumer become competitive. Researchers of behavioural intention, especially the ones who dwell on value theory paradigm, argued that before consumers possess favourable or unfavourable behavioural intention, they would consider the costs and benefits of a product (Zeithaml 1988). Consumers use cognitive assessment in evaluating sacrifices needed to consume certain product (Dodds 2002). Moreover, marketing literature introduced the variable of perceived sacrifice to represent the result of consumer evaluation process toward the amount of sacrifice in consuming a product (Zeithaml 1988). Some researchers on behavioural intention then added that variable as initiation variable of behavioural intention (Shukla 2010; Sheu 2010; Ruiz et al. 2008; Dodds 1991; Monroe and Chapman 1987), including in the context of land transportation (Jen and Hu 2003; Wen et al. 2005).

As explained before, behavioural intention is part of a chain that involves some antecedent variables. This article has already recognized three initiation variables, namely service quality, perceived sacrifice and image. Beside those three variables, based on study on previous researches, we have also identified some antecedent variables that mediate the relationship between initiation variables and behavioural intention, which are satisfaction (Molinari et al. 2008; Dahiyat et al. 2011; Kuo et al. 2011; Chuang and Cheng 2010; Lee and Lin 2005; Caruana 2002; Bloemer et al. 1998), trust (Aydin and Ozer 2005; Caceres and Paparoidamis 2007; Flavian and Guinaliu 2006), and perceived value (Lai and Chen 2010; Wen et al. 2005; Chang and Wang 2011; Tam, 2012). Based on above elaboration, this research expected that those six variables can explain behavioural intention of paratransit passengers. Descriptions of each variable and their relationship will be discussed in Sects. 2.4.1–2.4.6.

2.4.1 Service Quality

In service business, an important element that widely discussed in literature is service quality (Yusoff and Ismail 2008). Zeithaml (1988) defined quality as 'the consumer's judgement about a product [or service]'s overall excellence or superiority'. Further, marketing literature agreed that service quality show to what extent service performance matches consumers' expectations (Gronroos 1984,

1994; Parasuraman et al. 1985, 1988). If service performance matches or exceeds consumers' expectations, they will have favourable assessments toward service quality (Parasuraman et al. 1988).

In the context of public transports research, the definition of service quality in the marketing field is also adopted. For example, Transportation Research Board (1999, 2004) stated that 'service quality is a measure of how well the service level delivered matches customer expectations. Delivering quality service means conforming to customer expectations on a consistent basis [Lewis and Boom (1983 in Berry 1983)]'. Thus, the public transports passengers will have favourable perception of service quality if the performance matches or exceeds their expectations and the perception will be unfavourable if the service quality fall below the expectations.

Service marketing literature suggests that service quality is seen as multidimensional constructs (Berry et al. 1985; Parasuraman et al. 1985), but there is no consensus on the exact nature and measures of the constructs' dimensions (Brady and Cronin 2001). Nevertheless, the dimensions widely used by many researchers to measure the dimensions of service quality were identified by Parasuraman et al. (1985, 1988, 1991) and known as SERVQUAL (Parasuraman et al. 1988).

Parasuraman et al. (1985) found that the dimensions of service quality consist of 10 dimensions, namely reliability, responsiveness, competence, access, courtesy, communication, credibility, security, understanding/knowing the customer and tangibles. In their subsequent study, Parasuraman et al. (1988) summarized the 10 dimensions into five dimensions namely tangibles, reliability, responsiveness, assurance and empathy. The five dimensions had been tested and the results proved that these dimensions have high validity and reliability (Parasuraman et al. 1991). The five dimensions are known as SERVQUAL and the explanation can be seen in Table 2.1.

Although SERVQUAL has been applied by many researchers, the model of service quality is still criticized as described in the writings of Buttle (1996). For example, criticism of SERVQUAL is due to its generic nature. SERVQUAL is believed to be applicable to all types of industry (Ladhari 2009). In fact, some empirical studies suggest that the SERVQUAL dimensions are not stable in the context of a certain service industry (Carman 1990; Headley and Miller 1993; Mels et al. 1997; Durvasula et al. 1999). Furthermore, SERVQUAL dimensions only consider functional quality dimension, but not the technical quality dimension (Lagrosen et al. 2004). Therefore, many researchers are developing new models to fill the gap, such as Wen et al. 2005; Perez et al. 2007; Lai and Chen 2010.

The literature also shows that there are many researchers who identified and measured the dimensions of service quality for public land transports. Nevertheless, there is not yet a consensus among researchers about the dimensions of the public transport service quality. Some researchers adopted SERVQUAL dimensions, such as Gule (2009), Wijaya (2009), Perez et al. (2007), and Randheer et al. (2011). Meanwhile, other researchers proposed different dimensions. For example, Prasad and Shekhar (2010) used eight dimensions, namely tangibility, reliability, assurance, responsiveness, empathy, service product, social responsibility and

Table 2.1 Five dimensions of service quality

No	Dimensions	Explanation
1	Tangibles	'Physical facilities, equipment and appearance of personne'
2	Reliability	'Ability to perform the promised service dependably and accurately'
3	Responsiveness	'Willingness to help customers and provide prompt service'
4	Assurance	'Knowledge and courtesy of employees and their ability to inspire trust and coincidence'
5	Empathy	'Caring, individualized attention the firm provides its customers'

(Parasuraman et al. 1988)

Table 2.2 Service quality dimensions that are not taken from SERVQUAL

Author(s)	Dimensions	Transportation type	Country
Lai and Chen 2011	Core service and psychical environment	Mass rapid transit	Taiwan
Prasad and Shekhar 2010	Assurance, empathy, Reliability, responsiveness, tangibles, comfort, connection and convenience	Railways	India
Joewono and Kubota 2007	Availability, accessibility, reliability, information, customer service, comfort, safety and security, fare and environmental impact	Paratransits	Indonesia
Hu and Jen 2003	Interaction with passengers, tangible service equipment, convenience of service and operating management support	Bus service	Taiwan
Wen et al. 2005	Onboard amenity, crews attitude, station performance, operational performance	Bus service	Taiwan
Caro and Garcia 2007	Personal interaction, design, physical environment and outcome	Urgent transport service	Spain
Transportation Research Board (1999, 2004)	Safety, comfort, ease of using the system, convenience, performance/reliability, facilities and value	Train/bus	USA

service delivery. Lai and Chen (2011) proposed two dimensions of service quality which are core services and physical services. Table 2.2 shows the dimensions service quality that proposed by researchers who do not adopt the SERVQUAL.

2.4.2 *Perceived Value*

Value can be considered as the basis for all marketing activities (Holbrook 1994; Yang and Peterson 2004). In short, value is a customer's overall assessment of the usefulness of a product based on perceptions of what was gained and sacrificed

(Zeithaml 1988). Perceived value gives an overview of the customer's perception of a company in considering what they want, and believe that they get benefits from a product (Woodruff 1997). In line with the previous opinion, Slater and Narver (1994) said that value is the customer's perceived quality adjusted with the relative price of the product. Moreover, value can be used as an indicator of emotional bond formed between customers and manufacturers in which the customers found that the product provided an additional value for them (Butz and Goodstein 1996).

From some of the definitions and concepts of value above, we may conclude that 'value is a trade-off between what the customer receives (e.g. quality, benefits, worth, utilities) and what he or she gives up to acquire and use a product (e.g. price, sacrifices)'. Therefore, 'customer value is a customer's perceived preference for and evaluation of those product attributes, attribute performances and consequences arising from use that facilitate (or block) achieving the customer's goals and purposes in use situations' (Woodruff 1997).

In the context of public land transportation, perceived value is also understood as passengers' assessment on a comparison between the benefits and sacrifices (Lai and Chen 2010; Wen et al. 2005; Jen and Hu 2003). Thus, the passengers perceived value will increase if passengers' evaluation of the 'difference' between benefits and sacrifices yield positive result (meaning the benefits are higher than costs). Consequently, passengers perceived value will decrease if the passengers' assessment of the 'difference' between the benefits and cost generate negative result (meaning the costs outweigh the benefits).

2.4.3 Satisfaction

One of the most famous concepts that were built from various disciplines such as marketing, consumer research, psychology, economic wealth and economics is satisfaction (Akbar and Parvez 2009). Satisfaction has become popular topic in the marketing literature in the last 30 years (Gallo 2011). Many definitions had been proposed to operationalize this construct.

Woodruff (1993) stated that customer satisfaction tend to be treated as consumer attitudes in evaluating goods or services. This definition was confirmed by Bitner and Zeithaml (2003). They stated that satisfaction is 'customer evaluation of a product or service in terms of whether the product or service has met their needs and expectations'. In line with Bitner and Zeithaml, Day (1984 in Tse and Wilton 1988) also view satisfaction as 'the customer's response to the evaluation of the perceived discrepancy between prior expectations (or some other norm of performance) and actual performance of the product as perceived after its consumption'. Hunt (1977 in Letcher and Neves 2010) defined customer satisfaction as 'the favourableness of individual's subjective evaluation of the various outcomes and experiences associated with buying it or using it'. Based on these definitions, we can conclude that satisfaction is an assessment of customers' emotional affection on the overall performance of a product or service in fulfiling their expectations. Customers will be satisfied if the overall performance of the products meet or exceed expectations and vice versa (Kotler and Keller 2012).

This assessment (satisfaction) is holistic, covering all aspects related to public transport services—including not only the services, but also prices and situation (Zeithaml and Bitner 2003)—and is affected by the accumulated experience of the past, in contrast to service quality (Lai and Chen 2010). Service quality is a cognitive assessment, based on specific aspects of the service and it is transactional in nature (Parasuraman et al. 1988; Cronin and Taylor 1992). Given the conditions, it is not surprising that some empirical studies showed that service quality does not affect customer satisfaction (e.g. Oh 1999; Chen and Tsai 2007). Customers who have favourable perceived service quality are not necessarily satisfied since there are other factors which may upset them, such as prices or situation factor.

Researchers in the field of public land transport services also understand that passenger's satisfaction is the result of affective (emotional) evaluation on the overall performance of the public transport services in meeting their expectations (Lai and Chen 2010; Wen et al. 2005). Thus, passengers' will be satisfied if they emotionally feel that the overall performance of public transport services meet or exceed their expectations. In contrast, they will be dissatisfied if they emotionally feel that the overall performance of public transport services does not meet their expectations.

2.4.4 Trust

Relationship marketing literature has identified that trust is a variable that plays an important role in building relationships with customers (Berry 1995; Morgan and Hunt 1994; Palmatier et al. 2006). Furthermore, Wilson (1995) mentioned that 'trust is a fundamental relationship model building block and as such is included in most relationship models'. Literature also called trust under different names, such as trustworthiness, credibility, benevolence and honesty (Palmatier et al. 2006).

Some researchers have tried to propose definitions of trust. Although different, trust, generally, has been defined in one of the two possible ways: (1) as a willingness or intention to depend on the trustee (i.e. a trusting intention) and/or (2) as a confident belief or expectation (i.e. a trusting belief) (Dickey et al. 2007).

One of the researchers who support the first view of trust is Moorman et al. (1993). They defined trust as 'willingness to rely on an exchange partner in whom one has confidence' (Moorman et al. 1993). This definition was also used by Berry (1995), a marketing expert who is considered to have a major contribution to the birth of the field of relational marketing.

Morgan and Hunt (1994) criticized the first definition of trust, especially the definition of Moorman et al. (1993) above. Morgan and Hunt (1994) stated that the willingness or intention to depend on the trustee is the outcome of trust and not the part of the definition of trust as behavioural intention is considered as the outcome of attitude, but not the part of the definition of attitude itself. Given that, Morgan and Hunt (1994) stated that trust is 'existing when one party confidence in the exchange partner's reliability and integrity'. The view of Morgan and Hunt (1994) on trust was followed by other relational marketing researchers, such as Palmatier et al. (2006) and Aurier and N'Goala (2010).

Palmatier et al. (2006) explicitly stated that the definition of trust is 'confidence in an exchange partner's reliability and integrity'. Thus, customer's trust is customer's confidence in the service provider/brand's reliability and integrity. Boon and Holmes (1991) stated that trusting beliefs involve perceptions that the other party will act in ways favourable to the trustor. Furthermore, Hagen and Soonkyoo (1998) found that perceptions involve trusting beliefs that the other party has ethical, efficacious or favourable characteristics. Although researchers expressed trust in different ways, in general, they view trust as a confident belief. Furthermore, in general, it can be stated that trust is the confidence that other party will behave in a favourable or positive characteristics.

The literatures on public land transport service are also looking at trust as a confident belief (Wen et al. 2005). Given this, passengers' trust will increase if they believe that public transport will act in favourable and positive characteristics. In contrast, trust will decline if the passenger's do not believe the integrity and reliability of public transports.

2.4.5 *Image*

Image is one of the most important variables in the world of marketing communications (Shimp 2010). Image is an indicator used to measure the effectiveness of an integrated marketing communications program (Shimp 2010). Furthermore, service marketing literature identified image as an important factor in the overall evaluation of service (Grönroos 1984; Gummesson and Grönroos 1988).

Many experts tried to examine image and diverse definitions identified in the literature. Below are partial definitions of image contained in the marketing literature:

- The global impression (overall impression) that is in the minds of consumers (Zimmer and Golden 1988)
- Brand image is 'perceptions about a brand as reflected by the brand associations held in consumer memory' (Keller 1993)
- 'Associations and meanings connected with an organization' (Balmer 1998)
- 'The net result of all the experiences, impressions, beliefs, feelings and knowledge that people have about a company' (Aaker 1996 in Bridson and Mavondo 2011).

Although scholars defined image in different ways, in general, it can be concluded that image is consumer's mental picture (Dobni and Zinkhan 1990) of the offering which includes symbolic meanings consumers associate with a specific product or service (Padgett and Allen 1997). Based on above definition, the image functions 'to define the product for consumers and differentiate the firm's offering' among competitors (Padgett and Allen 1997). Thus, in the context of public land transport services, image is passengers' mental picture of the offering which includes symbolic meanings consumers associate with a specific land transport service. Passengers will have positive image of public land transport if they have positive mental picture toward public land transport. On other hand, passengers with negative mental picture toward public land transport will have negative image of public land transport.

According to Markus (1977), image is believed to have similar characteristics to the self-scheme in which humans develop the knowledge system to interpret their perceptions of a company/product. In other words, image formed in the mind of the consumer through a procedure where information was processed and prepared in a sense on the basis of stored category (Kosslyn 1975; MacInnis and Linda 1987). Furthermore, Andreassen and Lindestad (1998) stated that image can be formed from the consumption experience from time to time or from information supplied by companies like advertising or direct sales.

Marketing literature has identified conflict related to whether the image affects consumer evaluations of the consumption or the other way around. For example, Selnes (1993) looked at image as the dependent variable of perceived service quality and customer satisfaction while Andreassen and Lindestad (1998) argued that the image is an independent variable and proposed that image has a positive influence on customer perception of the perceived service quality and customer satisfaction.

The first view, which assumes that consumer evaluation affects image, is acceptable for consumption context that based on functional utilitarian need. In that particular context, consumers focused on aspects of how the product is capable of fulfilling its functions. Therefore, the central part of the brain, the ventral putamen, will be activated so that the evaluation results will be objective because the consumers' evaluation will only based on the experience of consumption (Shimp 2010).

On the other hand, in the context of consumption based on hedonic experiential needs, other parts of the brain, namely the medial cortex prefrontal will be activated if there are clues that indicate hedonic aspects, such as image (Shimp 2010; Solomon 2012). This will make the consumer's evaluation of a product biased. Consumers will evaluate the positive experience of consumption of products if there is information indicating that the product is in line with the positive things that he believed (Solomon 2012). In this context, the product information will be evaluated as a single integrated part with consumption experience (Wilcox et al. 2011). Thus, the image will affect consumer evaluations on consumption and not the other way around. Furthermore, according to Grönroos (1988 in Andreassen and Lindestad 1998), image plays as a filter that affects the perception of the company's operations, where it can form the 'expected quality' of a consumer.

Empirically, this has been proven by a study which evaluates consumer ratings of Cola products (Coca Cola and Pepsi). Coca Cola is known to have a better image than Pepsi. At the time of blind tests, in which participants were not informed which Cola brand products they consume, the result showed that the ventral putamen was activated and consumers chose Pepsi rather than Coca Cola because Pepsi taste better than Coca Cola. However, when consumers were told Cola brand products they were going to consume, results showed that the medial prefontral cortex was activated and consumers chose Coca Cola instead of Pepsi (Shimp 2010; Solomon 2012).

As described at the beginning of Sect. 2.4, 'means of transport' meet not only functional utilitarian needs, but also hedonic experiential needs. Furthermore, even marketing literature identified automotive products are more likely to meet hedonic experiential needs rather than functional utilitarian needs (Earl 2012). This is also reinforced by transportation literature that shows that growth of private vehicles always in harmony with the growth of income (Cox 2010). This condition indicates that means of transport has become part of a lifestyle than a person's mobility needs. With that in mind, this study adopts the view that image affects consumer evaluations on consumption and not the other way around.

2.4.6 Perceived Sacrifice

The concept of perceived sacrifice stems from an understanding about the importance of price paid by the consumer. Furthermore, pricing literature mentioned that the willingness to pay depends not only on the value obtained but also the price that consumers should pay (Nagle and Hogan 2006). In general, prices are often seen as the exchange value of a product or service. More clearly, price is often defined by the number of the price paid by the consumer for a product or service. In the academic world, the early outlook on prices is also oriented on the money actually incurred by a person or a monetary price. However, this view was then refuted firmly by Zeithaml (1988) with his means-end model. He revealed that the price paid by the consumer is not only monetary price, but also non-monetary price such as time, searches, efforts and psychological burden. He declared that price is 'what is given up or sacrificed to obtain a product [or service]' (Zeithaml 1988). Thus, the concept of prices can also be referred to as perceived sacrifice.

Researchers in the field of public land transport passengers' behavioural intention are also considered that sacrifice includes not only monetary sacrifice but also non-monetary sacrifice (Wen et al. 2005; Lai and Chen 2011). Thus, they have identified the importance of perceived sacrifice. They argued that perceived sacrifice has indirect effect on behavioural intention through perceived value (Wen et al. 2005; Lai and Chen 2011).

Following Zeithaml (1988), in general, researchers on the field of public transport passengers' behavioural intention defined perceived sacrifice as both monetary and non-monetary sacrifices which must be paid by passengers to obtain public transport services (Wen et al. 2005; Lai and Chen 2010). Passengers of public transport will have higher perceived sacrifice [have unfavourable perceived sacrifice] if they have to do high sacrifices in order to get the service.

References

Aaker, D. A. (1996). *Building Strong Brands*, Free Press, New York, NY, in Bridson, Karrie and Mavando, Felix (2011). Corporate image in the leisure services sector. *Journal of Services Marketing, 25* (3), 190–201.

Ajzen, I. (1991). The theory of planned behavior. *Organizational Behavior and Human Decision Processes, 50*, 179–211.

Akbar, M. M., & Parvez, N. (2009). Impact of service quality, trust, and customer satisfaction on customer loyalty. *ABAC Journal, 29*(1), 24–38.

Al-Rousan, Ramzi, M., & Mohamed, B. (2010). Customer loyalty and the impacts of service quality: The case of five star hotels in Jordan. *International Journal of Human and Social Sciences, 5*(13), 886–892.

Andreassen, T. W., & Lindestad, B. (1998). Customer loyalty and complex service: The impact of corporate image on quality, customer satisfaction and loyalty for customers with varying degree of service expertise. *International Journal of Service Industry Management, 9*, 7–23.

Aurier, P., & N'Goala, G. (2010). The differing and mediating roles of trust and relationship commitment in service relationship maintenance and development. *Journal of the Academy of Marketing Science, 38*(3), 303–325.

Aurier, P., & Lanauze, G. S. D. (2011). Impacts of in-store manufacturer brand expression on perceived value, relationship quality and attitudinal loyalty. *International Journal of Retail & Distribution Management, 39*(11), 810–835.

Aurier, P., & Lanauze, G. S. D. (2012). Impacts of perceived brand relationship orientation on attitudinal loyalty: An application to strong brands in the packaged goods sector. *European Journal of Marketing, 46*(11), 1602–1627.

Aydin, S., & Ozer, G. (2005). The analysis of antecedents of customer loyalty in the Turkish mobile telecommunication market. *European Journal of Marketing, 39*(7), 910–925.

Bagozzi, R. P. (1995). Reflections on relationship marketing in consumer markets. *Journal of the Academy of Marketing Science, 23*(4), 272–277.

Balmer, J. (1998). Corporate identity and the advent of corporate marketing. *Journal of Marketing Management, 14*, 963–996.

Berry, L. L. (1983). Relationship marketing. In L. L. Berry, G. L. Shostack & G. D. Upah (Eds.), *Perspectives in services marketing*, Chicago, IL: American Marketing Association.

Berry, L. L. (1995). Relationship marketing of services—growing interest, emerging perspectives. *Journal of the Academy of Marketing Science, 23*(4), 236–245.

Berry, L. L., Zeithaml, V. A., & Parasuraman, A. (1985). Quality counts in services, too. *Business Horizon*, 44–52.

Bloemer, J., de Ruyter, K., & Wetzels, M. (1999). Linking perceived service quality and service loyalty: A multi-dimensional perspective. *European Journal of Marketing, 33*(11/12), 1082–1106.

Bloemer, J., de Ruyter, K., & Peeters, P. (1998). Investigating drivers of bank loyalty: the complex relationship between image, service quality and satisfaction. *International Journal of Bank Marketing, 16*(7), 276–286.

Bloemer, Josee, & de Ruyter, K. (1998). On the relationship between store image, store satisfaction and store loyalty. *European Journal of Marketing, 32*(5/6), 499–513.

Boon, S. D., & Holmes, J. G. (1991). The dynamics of interpersonal trust: Resolving uncertainty in the face of risk in Colquitt, Jason A., Brent A., Scott, and Jeffery A., LePine (2007). Trust, trustworthiness, and trust propensity: A meta-analytic test of their unique relationships with risk taking and job performance. *Journal of Applied Psychology, 92*(4), 909–927.

Brady, M. K., & Cronin, J. J. (2001). Some new thoughts on conceptualizing perceived service quality: A hierarchical approach. *Journal of marketing, 65*, 34–49.

Bridson, K., & Mavando, F. (2011). Corporate image in the leisure services sector. *Journal of Services Marketing, 25*(3), 190–201.

Bruhn, M. (2003). *Relationship marketing management of customer relationship*. Prentice Hall: Pearson Education Limited Edinburgh Gate Harlow Exxex CM202JE England, UK.

Bush, A. J., Martin, C. A., & Bush, V. D. (2004). Sports celebrity influence on the behavioral intentions of generation Y. *Journal of Advertising Resaeach,* 108–118. DOI: 10.1017/S0021849904040206.

Buttle, F. (1996). SERVQUAL: Review, critique, research agenda. *European Journal of Marketing*, *30*(1), 8–32.

Butz, H. E., Goodstein Jr., & L. D. (1996). Measuring customer value: Gaining the strategic advantage. *Organization Dynamics*, (Winter), 63–77.

Caceres, R. C., & Paparoidamis, N. G. (2007). Service quality, relationship satisfaction, trust, commitment and business-to-business loyalty. *European Journal of Marketing, 41*(7), 836–867.

Carman, J. M. (1990). Consumer perceptions of service quality: An assessment of the SERVQUAL dimensions. *Journal of Retailing, 66*(Spring), 33–55.

Caro, L. M., & Gracia, J. A. M. (2007). Measuring perceived service quality in urgent transport service. *Journal of Retailing and Consumer Services, 14*, 60–72.

Caruana, A. (2002). Service loyalty—The effects of service quality and the mediating role of customer satisfaction. *Europan Journal of Marketing, 36*(7/8), 811–828.

Chang, H. H., & Wang, H.-W. (2011). The moderating effect of customer perceived value on online shopping behaviour. *Online Information Review, 35*(3), 333–359.

Chen, C.-F., & Tsai, D. C. (2007). How destination image and evaluative factors affect behavioral intentions? *Tourism Management, 28*, 1115–1122.

Chiou, J. S., & Shen, C. C. (2006). The effects of satisfaction, opportunism, and asset specificity on consumers' loyalty intention toward internet portal sites. *International Journal of Service Industry Management, 17*(1), 7–22.

Chiou, J. S., Wu, L. T., & Sung, Y. P. (2009). Buyer satisfaction and loyalty intention in online auctions: online auction web site versus online auction seller. *Journal of Service Management, 20*(5), 521–543.

Chuang, C.-F., & Cheng, C.-J. (2010). A study of institutional repository service quality and users' loyalty to college libraries in Taiwan: The mediating and moderating effects. *Journal of Convergence Information Technology, 5*, 8.

Clemes, M. D., Gan, C., Kao, T. H., & Choong, M. (2008). An empirical analysis of customer satisfaction in international air travel. *Innovative Marketing*, *4*, 50–62.

Cox, P. (2010). *Moving people, sustainable transport development*. Cape Town: UCT Press.

Cronin, J. J., & Taylor, S. A. (1992). Measuring service quality: A reexamination and extension. *Journal of Marketing, 56*, 55–68.

Cronin, J. J., Brady, M. K., & Hult, G. T. M. (2000). Assessing the effects of quality, value, and customer satisfaction on consumer behavioral intentions in service environments. *Journal of Retailing*, *76*(2), 193–218, N: 0022–4359.

Dahiyat, S. E., Akroush, M. N., & Abu-Lai, B. N. (2011). An integrated model of perceived service quality and customer loyalty: an empirical examination of the mediation effects of customer satisfaction and customer trust. *International Journal of Services and Operations Management, 9*(4), 453–490.

Day, R. L. (1984). Modeling choices among alternative responses to Dissatisfaction. *Advances in Consumer Research, 11*.In T. C. Kinnear (Ed.), Ann Arbor, MI: Association for Consumer Research, 469–479. In Tse D. K & Wilton P. C (Eds) (1988). Models of consumer satisfaction formation: An extension. *Journal of Marketing Research, XXV*, 2004–2012.

Dickey, M. H., McKnight, D. H., & George, J. F. (2007). The role of trust in franchise organizations. *International Journal of Organizational Analysis, 15*(3), 251–282.

Dobni, D., & Zinkhan, G. M. (1990). In search of brand image: A foundation analysis. *Advances in Consumer Research, 17*(1), 110–119.

Dodds, W. B. (1991). In search of value: how price & store name information influence buyer's product perceptions. *The Journal of Service Marketing*, *5*(3), 27–36.

Dodds, W. B. (2002). The effects of perceived and objective market cues on consumers' product evaluations. *Marketing Bulletin, 13*, 1–14.
Durvasula, S., Lysonski, S., & Mehta, S. C. (1999). Testing the SERVQUAL scale in the business-to-business sector: The case of ocean freight shipping service. *Journal of Services Marketing, 13*(2), 132–148.
Earl, P. E. (2012). Experiential analysis of automotive consumption. *Journal of Business Research, 65*, 1067–1072.
Egan, J. (2011). *Relationship marketing* (4th ed.). Harlow: Pearson Educational Limited.
Engel, J. F., Roger, D. B., & Miniard, P. (1994). *Consumer behaviour* (6th ed.). Translated by Budiyanto FX. *Perilaku Konsumen*. Jilid 1 dan 2. Jakarta: Binarupa Aksara.
Falk, T., Hammerschmidt, M., & Schepers, J. J. L. (2010). The service quality satisfaction link revisited: exploring a symmetries and dynamics. *Journal of the Academy of Marketing Science, 38*(3), 288–302.
Flavián, C., & Guinalíu, M. (2006). Consumer trust, perceived security and privacy policy: Three basic elements of loyalty to a web site. *Industrial Management & Data Systems, 106*(5), 601–620.
Fornell, C. (1992). A national customer satisfaction barometer: The Swedish experience. *Journal of Marketing, 5*, 1–21.
Gallo, M. (2011). Measuring passenger satisfaction: A strategy based on rasch analysis and the anom. *Journal of Applied Quantitative Methods, 6*(2), 27–35.
Georgescu-Roegen, N. (1936). The pure theory of consumer's behavior. *Quarterly Journal of Economics, 50*, 545–593.
Gowdy, J., & Mesner, S. (1998). The evolution of Georgescu-Roegen's bioeconomics. *Review of Social Economy, LVI*, 136–156.
Grönroos, C. (1984). A Service Quality Model and Its Marketing Implications. *European Journal of Marketing, 18*(4), 36–45.
Grönroos, C. (1988). Service quality: the six criteria of good perceived service quality. Review of Business, 9, Winter, 10–13, in Andersen, T.W. & B. Lindestad. (1998). Customer loyalty and complex service: the impact of corporate image on quality, customer satisfaction and loyalty for customers with varying degree of service expertise. *International Journal of Service Industry Management, 9*, 7–23.
Gronroos, C. (1994). From marketing mix to relationship marketing: Towards a paradigm shift in marketing. *Management Decision, 32*(2), 4–20.
Gule, X. M. (2009). Improving the service quality of taxi operators in the Nelson Mandela bay. Thesis, Faculty of Business & Economic Sciences, Nelson Mandela Metropolitan University. Retrieved from http://dspace.nmmu.ac.za:8080/jspui/handle/10948/1115
Gummesson, E., & Grönroos, C. (1988). Quality of services: Lessons from the Product Sector. *Add Value to Your Service. C. Suprenant. Chicago*, American Marketing Association.
Gummesson, E. (2008). *Total relationship marketing* (3rd ed.). Oxford, USA: Butterworth-Heinemann.
Hagen, J. M., & Soonkyoo, C. (1998). Trust in Japanese interfirm relations: Institutional sanctions matter. *Academy of Management Review, 23*(3), 589–601.
Headley, D. E., & Miller, S. J. (1993). Measuring service quality and its relationship to future consumer behavior. *Journal of Health Care Marketing, 13*(4), 32–41.
Holbrook, M. B. (1994). The nature of customer's value: An axiology of service in consumption experience. In R. T. Rust, & R. L. Oliver (Eds.), *Service quality: New directions in theory and practice* (pp. 21–71). Thousand Oaks, CA: Sage Publications.
Huang, Y-K. (2009). The effect of airline service quality on passengers' behavioural intentions using SERVQUAL scores: A TAIWAN case study. *Journal of the Eastern Asia Society for Transportation Studies*, 8.
Hunt, K. H. (1977). CS/D—Overview and future directions. In D. W. Letcher & J. S. Neves (Eds.), *Determinant of undergraduate business student satisfaction, Research in Higher Education Journal, 01–26 (2010).*

Jaiswal, A. K., & Niraj, R. (2011). Examining mediating role of attitudinal loyalty and nonlinear effects in satisfaction-behavioral intentions relationship. *Journal of Services Marketing, 25*(3), 165–175.

Jen, W., & Hu, K. C. (2003). Application of perceived value model to identify factors affecting passengers' repurchases intention on city bus: a case of the Taipei metropolitan area. *Transportation, 30*, 307–327.

Joewono, T. B., & Kubota, H. (2007). User satisfaction with paratransit in competition with motorization in Indonesia: anticipation of future implications. *Transportation, 34*, 337–354.

Kandampully, J., & Hu, H.-H. (2007). Do hoteliers need to manage image to retain loyal customers? *International Journal of Contemporary Hospitality Management, 19*(6), 435–443.

Keller, K. (1993). Conceptualizing, measuring, and managing customer based equity. *Journal of Marketing, 57*, 1–22.

Kosslyn, S. M. (1975). Information representation in visual images. *Cognitive Psychology, 7*, 341–370.

Kotler, P., & Keller, K. L. (2012). *Marketing Management* (14th ed.). England: Pearson.

Kucukaksoy, I. (2011). Adam Smith's conceptual contributions to international economics: Based on the wealth of nations. *Business & Economic Horizons, 4*(1), 108–119.

Kuo, N. T., Chang, K. C., Cheng Y. S., & Lai, C. H. (2011). The impact of service quality, customer satisfaction and loyality in the restautant industry: Moderating effect of perceived value. *Quality & Reliability ICQR—IEEE International Conference*, pp. 551–555.

Ladhari, R. (2009). A review of twenty years of SERVQUAL research. *International Journal of Quality and Service Sciences, 1*(2), 172–198.

Lagrosen, S., Seyye-Hashemi, R., & Leitner, M. (2004). Examination of the dimensions of quality in higher education. *Quality Assurance in Education, 12*(2), 61–69.

Lai, W.-T., & Chen, C.-F. (2011). Behavioral intention of public transit passenger—the role of service quality, perceived value, satisfaction and involvement. *Transport Policy, 18*, 318–325.

Letcher, D. W., & Neves J. S. (2010). Determinant of undergraduate business student satisfaction. *Research in Higher Education Journal*, 01–26.

Lee, G.-G., & Lin, H.-F. (2005). Customer perceptions of e service quality in online shopping. *International Journal of Retail & Distribution Management, 33*(2), 161–176.

Lien, T. B., & Yu, C. C. (2001). An Integrated Model for the effects of perceived product, perceived service quality, and perceived price fairness on customer satisfaction and loyalty. *Journal of Consumer Satisfaction, Disatisfaction, & Complaining Behaviour, 14*, 125–140.

Liu, C., Marchewka, J. T., Lu, J., & Yu, C.-S. (2005). Beyond concern—a privacy-trust-behavioral intention. *Information & Management, 42*, 289–304.

Lynd, R. S. (1936). Democracy's third estate: The consumer. *Political Science Quarterly, 51*(4), 481 515.

MacInnis, D., & Linda, P. (1987). The role of imagery in information processing: Review and extensions. *Journal of Consumer Research, 13*, 473–491.

Markus, H. (1977). Self-schemas and processing information about the self. *Journal of Personality and Social Psychology, 35*, 63–78.

Mels, G., Boshoff, C., & Nel, D. (1997). The dimensions of service quality: The original European perspective revisited. *Services Industries Journal, 17*(1), 173–189.

Molinari, L. K., Abratt, R., & Dion, P. (2008). Satisfaction, quality and effects on repurchase and positive word-of-mouth behavioral intention in a B2B services context. *Journal of Services Marketing, 22*(5), 363–373.

Monroe, K. B., & Chapman, J. D. (1987). Framing effects on buyers' subjective product evaluations. *Advances in Consumer Research, 14*, 193–197.

Moorman, C., Rohit, D., & Zaltman, G. (1993). Factors affecting trust in market research relationships. *Journal of Marketing, 57*, 81–101.

Morgan, R. M, & Hunt, S. D. (1994). The commitment-trust theory of relationship marketing. *Journal of Marketing*, 58(3), 20–38. Retrieved from http://dx.doi.org/10.2307/1252308

Mowen, J. C, dan Minor, M. (2002). *Consumer Behaviour* (5th ed). Translated by Dwi Kartini Yahya. *Perilaku Konsumen*. Jilid 1 dan 2. Jakarta : Erlangga.

Nagle, T. T. & Hogan, J. (2006). *The strategy and tactics of pricing: A guide to growing more profitably*. Prentice Hall.

Nguyen, Nh, & Leblanc, G. (1998). The mediating role of corporate image on customers' retention decisions: An investigation in financial services. *International Journal of Bank Marketing, 16*(2), 52–65.

Oh, H. (1999). Service quality, customer satisfaction, A customer value: A holistic perspective. *Hospitality Management, 18*, 67–82.

Olsen, S. O., Tudoran, A. A., Brunso, K., & Verbeke, W. (2013). Extending the prevalent consumer loyalty modelling: The role of habit strength. *European Journal of Marketing, 47*(1–2), 303–323.

Padgett, D., & Allen, D. (1997). Communicating experiences: A narrative approach to creating service brand image. *Journal of Advertising, 26*(4), 49–62.

Palmatier, R. W., Rajiv, P., Dant, D. G., & Evans, K. (2006). Factors influencing the effectiveness of relationship marketing: A meta-analysis. *Journal of Marketing, 70*, 136–153.

Parasuraman, A., Berry, L. L., & Zeithaml, V. A. (1991). Refinement and reassessment of the SERVQUAL scale. *Journal of Retailing, 67*(4), 420–450.

Parasuraman, A., Zeithaml, V., & Berry, L. (1985). A conceptual model of service quality and its implications for future research. *Journal of Marketing, 49*, 41–50.

Parasuraman, A., Zeithaml, V., & Berry, L. (1988). SERVQUAL: A multiple item scale for measuring customer perceptions of service quality. *Journal of Retailing, 64*, 12–40.

Pérez, M. S., Abad, J. C. G., Carrillo, G. M. M., & Fernández, R. S. (2007). Effects of service quality dimensions on behavioural purchase intentions: A study in public-sector transport. *Managing Service Quality, 17*(2), 134–151.

Peterson, R. A. (1995). Relationship marketing and the consumer. *Journal of the Academy of Marketing Science, 23*(4), 278–281.

Prasad, M. D., & Shekhar, B. R. (2010). Development of railqual : A service quality scale for measuring Indian railway passenger services. *Management Science & Engineering, 4*(3), 87–94.

Pugliese, M. J., & Cagan, J. (2002). Capturing a rebel: Modeling the Harley-Davidson brand through a motorcycle shape grammar. *Research in Engineering Design, 13*, 139–156.

Randheer, K., AL-Motawa, A. A., & Vijay, P. I.. (2011). Measuring commuters' perception on service quality using SERVQUAL in public transportation. *International Journal of Marketing Studies*, 3(1), 21–24.

Ruiz, D. M., Gremler, D. D., Washburn, J. H., & Carrión, G. C. (2008). Service value revisited: specifying a higher-order, formative measure. *Journal of Business Research, 61*, 1278–1291.

Ryu, K., Han, H., & Kim, T.-H. (2008). The relationships among overall quick-casual restaurant image, perceived value, customer satisfaction, and behavioral intentions. *International Journal of Hospitality Management, 27*, 459–469.

Sanzo, M. J., Santos, M. L., Alvarez, L. I., & Vazquez, R. (2007). The effect of a buyer's market orientation on attitudinal loyalty toward a supplier: Is dependence a moderator? *Supply Chain Management: An International Journal, 12*(4), 267–283.

Schiffman, L. G., & Kanuk, L. L. (1994). *Consumer behavior* (5th ed.). London: Prentice Hall.

Selnes, F. (1993). An examination of the effect of product performance on brand reputation, satisfaction and loyalty. *European Journal of Marketing, 27*(9), 19–35.

Sheth, J. N., & Parvatiyar, A. (1995). Relationship marketing in consumer markets: antecedents and consequences. *Journal of the Academy of Marketing Science, 23*(4), 255–271.

Sheu, T.-S. (2010). Exploring the differential affections of service quality, sacrifice, perceived value, and customer satisfaction on university students' favorable and unfavorable behavioral intention. *Journal of Quality, 17*(6), 483–500.

Shimp, T. A. (2010). *Integrated marketing communication in advertising and promotion* (8th ed.). CENGAGE Learning: South-Western.

Shukla, P. (2010). Effects of perceived sacrifice, quality, value, and satisfaction on behavioral intention in the service environment. *Services Marketing Quarterly, 31*, 466–484.

Slater, S. F., & Narver, J. (1994). Market orientation, customer value, and superior performance. *Business Horizons, 37*, 22–28.

Solomon, M. R. (2012). *Consumer behavior: Buying, having and being* (10th ed.). London: Prentice Hall.

Sondoh, S. L, Jr, Omar, M. W., & Wahid, N. A. (2007). The effect of brand image on overall satisfaction and loyalty intention in the context of color cosmetic. *Asian Academy of Management Journal, 12*(1), 83–107.

Tam, J, L.-M. (2012). The moderating role of perceived risk in loyalty intentions: an investigation in a service context. *Marketing Intelligence & Planning, 30*(1), 33–52.

Transportation Research Board. (1999). *A handbook for measuring customer satisfaction and service quality*. Report: TRCP. 47.

Transportation Research Board. (2004). Transit capacity and quality of service manual. *TRCP Report* 100 (2nd ed).

Wen, C.-H., Lan, L. W., & Cheng, H.-L. (2005). Structural equation modelling to determine passenger loyalty toward intercity bus services. *Journal of the Transportation Research Board, 1927*, 249–255.

Wijaya, D. H. (2009). Study of service quality in the public bus transport: Customer complaint handling and service standards design—case study: Transjakarta busway & Värmlandstrafik ab bus. Thesis, Faculty of Economic Sciences, Communication & IT, Karlstads University.

Wilcox, K., Roggenveen, A. L., & Grewal, D. (2011). Shall I tell you now or later? Assimilation and contrast in the evaluation of experiential products. *Journal of Consumer Research, 38*(4), 763–773.

Wilson, D. T. (1995). An integrated model of buyer-seller relationships. *Journal of the Academy of Marketing Science, 23*(4), 335–345.

Woodruff, R. B. (1997). Customer value: The next source for competitive advantage. *Journal of the Academic Market Science, 25*(2), 139–153.

Woodruff, R. B. (1993). Developing and applying consumer satisfaction research: Implications for future research. *Journal of Consumer Satisfaction, Disatisfaction & Complaining Behavior, 6*, 1–11.

Yang, Z., & Peterson, R. T. (2004). Customer satisfaction, perceived value, and loyalty: The role of switching costs. *Psychology & Marketing, 21*(10), 799–822.

Yen, C. H & Lu, P. H. (2008). Effects of e-service quality on loyalty intention: An empirical study in online auction *Managing Service Quality, 18*(2), 127–146.

Yusoff, W. Z. W., dan Ismail, M., (2008). FM-SERVQUAL: A new approach of service quality measurement framework in local authorities. *Journal of Corporate Real Estate, 10*(2), 49–64.

Zeithaml, V., & Bitner, M. J. (2003). *Services marketing* (3rd ed.). Boston, MA: McGraw-Hill Irwin.

Zeithaml, V. A. (1988). Consumer perceptions of price, quality and value: A means and model and synthesis of evidence. *Journal of Marketing, 52*, 2–22.

Zeithaml, V. A., Berry, L. L., & Parasuraman, A. (1996). The behavioral consequences of service quality. *Journal of Marketing, 60*, 31–46.

Zhou, Z., Zhang, Q., Su, C., & Zhou, N. (2012). How do brand communities generate brand relationships? Intermediate mechanisms. *Journal of Business Research, 65*, 890–895.

Zimmer, M. R., & Golden, L. L. (1988). Impressions of retail stores: A content analysis of consumer images. *Journal Retail, 64*, 265–291.

Zins, A. H. (2001). Relative attitudes and commitment in customer loyalty models—Some experiences in the commercial airline industry. *International Journal of Service Industry Management, 12*(3), 269–294.

Chapter 3
Conceptual Model and Hypotheses

Abstract This chapter presents conceptual model and hypotheses which will be tested later. Chapter three also explains the arguments used for building the model and its hypotheses from both theoretical and empirical sides.

Keywords Conceptual model · Hypotheses

3.1 Conceptual Model

Conceptual model describes the relationship between variables investigated in the study (Sekaran and Bougie 2011). Mayer and Greenwood (1980) stated that the conceptual model is a model that describes the cause-effect relationship (causal orientation) between variables in order to explain a problem. Thus, the conceptual model of behavioural intention passenger public land transport is a model that describes the cause-effect relationship (causal orientation) between antecedent variables in order to explain the development of the passengers' behavioural intention.

In developing the passenger behavioural intention conceptual model, we see "public land transport" as one of "means of transport". Hence, we argue that the passenger behavioural intention of public land transport is also affected by variables that influence the behavioural intention of the consumers of others land "means of transports", such as motorcycle or cars.

The conceptual model of this study can be seen in Fig. 3.1. The conceptual model proposes that passengers' behavioural intentions are directly affected by 5 variables: image, service quality, value, satisfaction, and trust. Trust is influenced by 4 variables: image, service quality, value, and satisfaction. Satisfaction is influenced by image, service quality, and value. Value is influenced by perceived sacrifice, image, and service quality. Service quality is affected by image.

In contrast to previous studies (Jen and Hu 2003; Wen et al. 2005; Lai and Chen 2011), instead of using service quality (representing functional utilitarian need), this study chooses image (representing hedonic experiential needs) as the initiation

S. Sumaedi et al., *Public Transport Passengers' Behavioural Intentions*, SpringerBriefs in Business, DOI: 10.1007/978-981-4585-24-8_3,

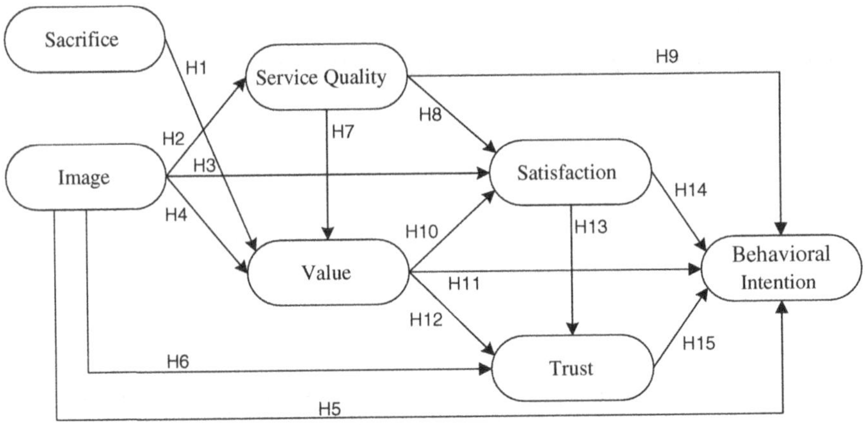

Fig. 3.1 Conceptual model

variable. It is based on the idea that human needs have shifted from functional to hedonic requirements (Falk et al. 2010; Aurier and N'goala 2010; Solomon 2012). More specifically, marketing literature has confirmed that "means of transport" tend to be used by consumers as a tool to fulfil their hedonic needs, whereas "means of transport" has become part of them and define who they are (Zhou et al. 2012; Keaveney et al. 2012; Earl 2012). This is also indicated by transportation researcher, Cox (2010), which he stated that more complex transportation problems arise along with the growth of income. Income growth tends to require adjustment in lifestyle and even identity reposition. In this context, refers to the Self Completion Theory, "means of transport" are part of the efforts to reposition passengers' identities (Solomon 2012).

The above conditions also occur in the context of public land transport passengers in Jabodetabek area, which is a metropolitan city in Indonesia. People make their means of transport as their status symbol, not merely functional and instrumental tool for the ease of mobility. This is reinforced by the efforts of the manufacturers to offer variety of products that support the lifestyle of a particular segment of society. Furthermore, advertisements also tend to emphasize the aspects of hedonic experiential needs rather than functional utilitarian needs. Table 3.1 gives an example of some of the automotive products in Indonesia and their advertisings.

Meanwhile, in the context of the needs of people with hedonic experiential needs, customer evaluation of the functional aspect of products, such as service quality, will be biased. Consumers will evaluate the positive experience of consumption of products if there is information indicating that the product is in line with the positive things that they believe (Solomon 2012). This is because there is a part of the brain, namely the medial prefontral cortex, which will be activated if there is a lead that shows hedonic aspect like image (Shimp 2010; Solomon 2012). In this context, according to (Grönroos 1988 in Andreassen and Lindestad 1998), image acts as a filter that affects the perception about company's operations, where

Table 3.1 Automotive products advertisements that emphasize the fullfilment of hedonic needs

Product	Advertisement	
	Tagline	Explanation
Yamaha-Mio Fino	Loving Mio Fino! Fashionable Matic	This advertisement emphasizes the physical appearance aspects of motorcycle (shape and colour) that are fashionable for women riders There are three types offered by Mio Fino: Classic, Fashion, and Sport. This advertisement doesn't highlight the performance of the motor. Motor design that is considered fashionable is offered by the Yamaha Mio Fino. Mio Fino specifically targeting the female consumer market who is very concerned about the appearance. Mio Fino presents to meet the needs of women about transport equipment and style
Honda Scoopy	Scoopy. Uniquely Happy	According to the advertisement, Honda Scoopy can the make everyone feel happy, unique, and fun. Honda Scoopy physical appearance can make people who saw it fall in love. Honda Scoopy has many colors and different designs with other motor matic. These make the rider look more unique. By reading the tagline and watching ads of Honda Scoopy on TVC, efforts to meet the hedonic needs of the consumer very strong look
New Honda Beat 2012	New Honda Beat. Can't Stop The New Beat	The advertisement say that the presence of the new Honda Beat 2012 can make people's heart singing, more colorful, and fun. New Honda Beat 2012 also changes their life to become more beautiful because it comes with a new color, the right size, agility, and new striping. New Honda beat also more economical. The advertisement of New Honda Beat 2012 in the TVC emphasizes aspects the fun and does not emphasize key aspects of motorcycle performance
Honda Revo 2007	Revo. Your Style	This advertisement emphasizes the design of Honda Revo 2007 that is spectacular, sharp character lines of the body, sleek, sporty, and aerodynamic. This advertisement highlights the body of Honda Revo from some sides. Tagline "Revo. Your Style" and the advertisement of Honda Revo have same mission that is offer Honda Revo 2007 as a means of transportation as well as meet their hedonic needs (style)
KIA All New Picanto	Smart, Stylish, Fun	The advertisement said that the all new Picanto cars make proud the riders, make them cool, have many awards, make hangout more exciting, easy parking, economical, comfortable, spacious, and have a broad baggage. KIA All New Picanto tagline is "Smart, Stylish, Fun". This tagline shows that this car is a smart and stylish car. It can also make the riders happy. Associated with Kia All New Picanto advertisement on TVC, besides "smart" (economical, comfortable, easy parking, spacious, and have a broad baggage), this car also offers fulfillment of human hedonic needs such as: "style" and "fun" by making them proud and cool, giving them a lot of awards, making the rider more fun with All New Kia Picanto

(continued)

Table 3.1 (continued)

Product	Advertisement	
	Tagline	Explanation
Yamaha Mio Soul	True Style True Expression	This advertisement highlights just a physical appearance of Yamaha Mio Soul. The advertisement said that Yamaha Mio Soul has a new look striping and two tone colours. No words or writings which highlight other aspects such as: performance. Tagline "True Style True Expression" can be interpreted that the Yamaha Mio Soul is the right choice for the consumer who wants to be more stylish and more expressive. So, Yamaha Mio Soul cannot be separated from the style and expression of the consumer
Mitsubishi Mirage	Mirage. Rainbow Your Life	This advertisement highlights the ability of the Mitsubishi Mirage in making the owner happy. Three aspects are shown in this advertisement are (1) Mitsubishi Mirage comes with audio system so that the driver can play a song, (2) the size of a car that is not too large, and (3) Mitsubishi mirage have diverse colors. In accordance with the tagline "Mirage. Rainbow Your Life", Mirage exists to give pleasure and make the days of the user to be colored

it can form the "expected quality" of a consumer. Balance Theory also proposes that consumers will try to justify what he did in order to create harmony between symbols and his true-self (Solomon 2012). Therefore, customers who felt that the image of a product is in accordance with his image will give a positive evaluation of the product even if the product is rationally negative in value.

Empirically, this has been proven by a study which evaluates consumer ratings of Cola products (Coca Cola and Pepsi). Coca Cola is known to have a better image than Pepsi. At the time of blind tests, in which participants were not informed which Cola brand products they consume, the result showed that the ventral putamen was activated and consumers chose Pepsi rather than Coca Cola because Pepsi taste better than Coca Cola. However, when consumers were told Cola brand products they were going to consume, results showed that the medial prefontral cortex was activated and consumers chose Coca Cola instead of Pepsi (Shimp 2010; Solomon 2012). It is also supported by the writings of McClure et al. (2004) which stated that consumer evaluations over Coca Cola tend to be more favourable when they consumed it from a glass with a Coca Cola logo on it than when they consumed it from plain glass.

Similar conditions can also occur in the context of "means of transport". Earl (2012)'s auto ethnography research showed that consumers tend to think irrationally when it comes to automotive product. This shows that consumers' evaluation can be biased. In Indonesia, this condition can be seen in the case of Toyota Avanza and Daihatsu Xenia. Although both products are very similar, the customer believes that Avanza has better quality than Xenia because the image of Toyota is better than Daihatsu. Therefore, this study believes that image, which is not service quality, is the initiation variable of behavioural intention since image will affect service quality in the context of hedonic experiential consumption.

As has been described at the beginning of this section, the conceptual model of public transportation passenger behavioural intention consists of relationships between variables in order to explain behavioural intention of public transportation passengers. The next section, hypotheses, will expose the alleged relationship between the variables and their arguments.

3.2 Hypotheses

3.2.1 The Influence of Sacrifice on Perceived Value

As mentioned in the description in Sect. 2.4, service providers not only respond to customer needs in the form of service, but they also make consumers sacrifice in return of the service. In the context of public transport services, passengers make both monetary and non-monetary sacrifices to obtain public transport services (Jen and Hu 2003; Wen et al. 2005; Lai and Chen 2011).

Marketing literature has long identified that companies are trying to manage price (Price Mix) and network distribution (Place Mix) (Gronroos 1994), including market leaders like Coca Cola (Keegan and Green 2011; Albanese 2001), Harley Davidson (Kotler 2003), Marlboro (Keegan and Green 2011; Belch and Belch 2003) and Subaru (Randazzo 2006). Effective price and place mix can be useful, among other things, in terms of making competitive scheme of sacrifices (Kotler 2003). This indicates that marketing practitioners have long believed that perceived sacrifice will affect the purchasing behaviour of customers.

Theory of Planned Behaviour (TPB) proposed by Ajzen (1991) suggests that one may not exhibit or wish to exhibit certain behaviour even though it is considered as good behaviour because he or she has the perception that he or she is incapable of doing such behaviour. Furthermore, the perception is growing due to the occurrence frequency of the inhibitors (Armitage and Christian 2006). In this context, perceived sacrifice can be an indicator of purchasing behaviour inhibitor to the customer.

More clearly, in order to discuss the relationship between perceived sacrifice and purchase intention, Zeithaml (1988) proposed means-end model that suggests perceived sacrifice will affect purchase intention through perceived value, whereas perceived sacrifice negatively affects perceived value. Some researchers have also argued that perceived sacrifice influences behavioural intention through perceived value (Sharma et al. 2012; Sheu 2010). In the context of public transport passengers' behavioural intention research, previous researches have also found the influence of perceived sacrifice on behavioural intention through perceived value (Huang 2009; Wen et al. 2005; Jen and Lu 2003).

Empirical studies to examine the relationship of perceived sacrifice and perceived value have been carried out in various contexts. The results of these studies demonstrated that perceived sacrifice negatively affect perceived value (e.g. Shukla 2010; Sheu 2010; Zhu et al. 2009; Ruiz et al. 2008; Dodds 1991; Monroe and Chapman 1987). Empirical studies in the context of public land transport also showed similar results, perceived sacrifice negatively affects perceived value (Jen and Hu 2003; Wen et al. 2005). Thus, perception of higher sacrifices in order to gain public land transport service will result in lower perceived service value. A similar condition is also thought to occur in the context of paratransit. Therefore, the first hypothesis of this study is stated as follows:

H1: Perceived sacrifice negatively affects perceived value of paratransit passengers

3.2.2 The Influence of Image on Service Quality, Satisfaction, Perceived Value, Trust, and Behavioural Intention

TPB has identified that the attitude of a person toward certain behaviour is determined by behavioural belief, which consists of two main components, outcome belief and outcome evaluation (Armitage and Christian 2006). Outcome belief indicates a person's perception of possible outcomes resulting from certain behaviour, while outcome evaluation shows the value of each outcome (Armitage and Cristian 2006). Service Quality and Perceived Value can be classified as cognitive-based attitude while satisfaction is affective-based attitude (Solomon 2012). As mentioned in the Sect. 3.1 Conceptual Model, image plays an important role in the context of customers with hedonic-based consumption patterns. Grönroos (1988 in Andreassen and Lindestad 1998), argued that image is a filter that affects the perception about company's operations. It can form the function of expected quality of a consumer. In other words, image represents purchasing behavioural belief of a customer (Solomon 2012). Thus, based on TPB, the image affects service quality, satisfaction, and perceived value.

Consumer behaviour literature suggests that consumer perceptions of service quality in the context of a product can be biased if the information related to the product is given beforehand (Wilcox et al. 2011; Solomon 2012). This is due to the consumers assessment on their consumption result will unify with the emotional reaction that resulted from the information provided (Shimp 2010; Wilcox et al. 2011), whereas if the information is favourable then the assessment would become more positive and if the information is unfavourable then the assessment would be more negative (Wilcox et al. 2011).

The literature has identified that the image is information that can affect a person's perception of the quality of a product (Shimp 2010; Wilcox et al. 2011; Solomon 2012). More clearly, Andreassen and Lindestad (1998) proposed that the image is a predictor of service quality. One frequently cited empirical evidence to support this is a study of consumer ratings of products between Coca Cola and Pepsi Cola. This study shows that the image of Coca Cola is better than Pepsi and it positive affects consumers' evaluation of Coca Cola (Shimp 2010; Solomon 2012). Other researchers, McClure et al. (2004) also noted that consumer evaluations over Coca Cola tend to be more favourable when they consumed it from a glass with a Coca Cola logo on it than when they consumed it from plain glass. In other words, these studies confirm the positive effect on service quality image.

Until now, there are no studies on public land transport services that investigate the relationship between image and service quality (Wen et al. 2005). Nevertheless, based on the previous description and the facts in the case of Xenia-Avanza described in Sect. 3.2 Conceptual Model, this study suspects that image affects perceived service quality. The second hypothesis of this study is stated as follows.

H2: Image positively affects perceived service quality of paratransit passengers

Andreassen and Lindestad (1998) suggested that image has a positive influence on customer satisfaction. Satisfaction model of Arthur D. Litle, as quoted by Lovelock et al. (2004), also put image as a predictor of satisfaction, together with service quality, product quality, and perceived price. Empirically, Wilcox et al. (2011) found that the favourable information given before the consumption of a product will increase consumer affective outcomes assessment, which is another name for the satisfaction. In other words, the study of Wilcox et al. (2011) found a positive effect of product information on customer satisfaction of experiential product. Literature has identified that image is one of product information that is often used by customers (Wilcox et al. 2011; Solomon 2012) and since "means of transport" is an experiential product (Zhou et al. 2012; Earl 2012), we can predict that image has positive impact on passengers' satisfaction. Other studies have also found a positive effect of image on satisfaction (Andreassen and Lindestad 1998; Bloemer and de Ruyter 1998; Minkiewicz et al. 2011; Alves and Raposo 2010; Palacio et al. 2002; Chien-Hsiung 2011; Alireza et al. 2011), including in the context of public transport (Meng et al. 2011; Liou and Tsao 2010; Zins 2001). Given this, this study assumes the positive influence of image on passenger satisfaction. The third hypothesis of this study is stated as follows.

H3: Image positively affects paratransit passengers' satisfaction
Perceived value represents a customer's perception of the ratio between perceived benefits and perceived sacrifice (Zeithaml 1988). The literature has confirmed the positive influence of perceived benefits to perceived value (Alsheikh and Bojei 2012), including in the context of public land transport (Jen and Hu 2003). Benefit from the consumption of a product is due to the fulfilment of customer needs. Thus, the perceived benefit can be emotional and or functional/instrumental (Falk et al. 2010). Based on the opinion of Tuskej et al. (2013), product image will provide emotional benefits. Thus, one can say that image has a positive influence on perceived value. Strictly speaking, Andreassen and Lindestad (1998) argued that image has a positive influence on perceived value. This opinion was confirmed by several empirical studies, including Milfelner et al. (2011), Ryu et al. (2012), Shafeiha and Saeednia (2011), and Alireza et al. (2011). In the context of public transport, empirical studies also show a positive effect of image on perceived value (Meng et al. 2011; Liou and Tsao 2010; Zins 2001). Therefore, a similar relationship is also expected to occur in the context of paratransit. The fourth hypothesis of this study is stated as follows.

H4: Image positively affects perceived value of paratransit passengers
Self-completion theory proposed that someone will try to find "something" that can be used as a complement to their identity. This theory is used by marketers to offer products that can supplement one's self-concept (Solomon 2012). In other words, a product is used as a tool to meet the needs of hedonic experiential rather than functional utilitarian (Falk et al. 2010). If someone feels that a product has something in common with him, someone would feel emotionally bound by it (Tuskej et al. 2013). This will make a person have a loyalty feeling to the product, which is the root of favourable behavioural intention (Zhou et al. 2012). Furthermore, the ability of a product to represent "something" depends on its image (Shimp 2010). Based on these explanations, according to the self-completion theory, it can be concluded that image affects behavioural intention.

Studies by Hooley and Cook (1984) and Chung and Lee (2003) put forward the proposition that image has a positive influence on purchase intention. Image's positive influence on behavioural intention was evident in a study conducted by Brandão et al. (2011). Other studies also confirmed the positive effect of image on behavioural intention (Chen and Tsai 2007; Ryu et al. 2008), including in the context of public transport (Meng et al. 2011). Given this, the study assumes that image has a positive influence on behavioural intention in the context of paratransit. The fifth research hypothesis is stated as follows.

H5: Image positively affects paratransit passengers' behavioural intention
According to social penetration theory, it can be implied that the image has an influence on trust since image represents consumer knowledge of the product. On the other hand, consumer knowledge will affect openness, equivalence, and the possibility and acceptance of criticism, which are the other forms of trust (Bruhn 2003).

The theory widely used relational marketing, trust-Commitment Theory, proposed that trust is influenced by shared values. On the other hand, in the context of people with hedonic consumption patterns, shared value will occur if the product can represent them, as the slogan "you are what you consume" (Solomon 2012). Thus, based on the trust-Commitment theory, image of the product will also affect trust.

Empirically, study conducted by Martin and Camerero (2009) with the aim of knowing the effect of perceived risk on online purchases proved that the image (reputation of the company) has a positive effect on trust. Several empirical studies have also confirmed similar findings (Flavián et al. 2005; Afzal et al. 2010; Kanibir and Nart 2009; Ball et al. 2004. Therefore, this study suspects a positive relationship between image and trust. The sixth research hypothesis is stated as follows:

H6: Image positively affects paratransit passengers' trust

3.2.3 The Influence of Service Quality on Perceived Value, Satisfaction, and Behavioural Intention

The study of behavioural intention almost always involves service quality in its conceptual models, including in the context of public land transport modes (Jen and Hu 2003; Wen et al. 2005; Lai and Chen 2011). This is reasonable considering service quality is believed to affect behavioural intention directly and indirectly through perceived value and satisfaction.

The mean-end model by Zeithaml (1988) proposed that service quality has a positive effect on perceived value. Many researchers have tried to examine the relationship service quality and perceived value, in which they have confirmed the existence of positive impact (Dodds et al. 1991; Cronin et al. 2000; Oh 1999; Kuo et al. 2009; Alireza et al. 2011; Choi et al. 2004). In the context of public land transport services, Wen et al. (2005) confirmed that service quality positively affects perceived value. Lai and Chen (2011) also found similar findings in a study conducted in the context of Mass Rapid Transport passengers. Therefore, this study assumes that the service quality will positively affect perceived value. The seventh research hypothesis is stated as follows.

H7: Service quality positively affects perceived value of paratransit passengers

The relationship between service quality and satisfaction is a popular topic in the marketing literature (Sureshchandar et al. 2002). Many researchers have tried to test the effect of service quality on customer satisfaction, such as Cronin and Taylor (1992), Ostrowski et al. (1993), Stank et al. (1999), Butcher et al. (2001), Kassim and Abdullah (2010), and Hellier et al. (2003). They confirmed that service quality positively affects customer satisfaction. Similar associations were also found in the empirical studies in the context of public transport passengers (e.g. Wen et al. 2005; Joewono and Kubota 2007; Lai and Chen 2011). Based on the description, service quality is thought to have a positive impact on paratransit passenger satisfaction. Eighth research hypothesis is stated as follows.

H8: Service quality positively affects paratransit passengers' satisfaction

Study of the influence of service quality on behavioural intention has also been widely conducted. The researchers generally found that service quality has a positive influence on behavioural intention (e.g. Lien and Yu 2001; Al-Rousan et al. 2010, Kuruuzum et al. 2010; Udo et al. 2010; Ravichandran et al. 2010; Boulding et al. 1993; Park et al. 2005). In the context of public land transport services, empirical studies conducted by Lai and Chen (2011) also confirmed the positive relationship between service quality and behavioural intention. Therefore, this study assumes that service quality has a positive influence on behavioural intention of paratransit passengers. Ninth research hypothesis is stated as follows.

H9: Service quality positively affects paratransit passengers' behavioural intentions

3.2.4 The Influence of Perceived Value on Satisfaction, Trust and Behavioural Intention

In addition to service quality, perceived value and satisfaction are two constructs that are often involved in the conceptual model of behavioural intention (Kuo et al. 2009; Choi et al. 2004; Ha and Jang 2010). Empirical studies on the influence of perceived value on satisfaction have been conducted by the researchers and they found a positive relationship between the two constructs (Andreassen and Lindestad 1998; Hellier et al. 2003). Research in the context of public land transport services also confirmed the positive effect of perceived value on satisfaction. For example, Lai and Chen (2011) found that perceived value has positive effect of the mass rapid transport passengers' satisfaction. Similar findings were obtained by Wen et al. (2005) who conducted research in the context of intercity bus passengers. Based on these explanations, perceived value is also thought to positively influence satisfaction in the context of paratransit. Tenth research hypothesis is stated as follows.

H10: Perceived value positively affects paratransit passengers' satisfaction
Value theory proposed that customers form relationships in order to maximize the value obtained while in learning theory, through the principle of reinforcement, revealed that customers tend to be loyal when they received clear value (Bruhn 2003). Relationship marketing literature used these theories as a basis of argument for the influence of perceived value on behavioural intention.

Means end Zeithaml (1988) put forward the proposition that perceived value positively influence purchase intention. Researchers conducted empirical studies to examine the relationship between perceived value and behavioural intention and they found that perceived value positively influence purchase intention (Petrick and Backman 2002; Petrick 2004; Chen and Sai 2008; Meng et al. 2011). In the context of public land transport services, Lai and Chen (2011) also confirmed the positive association between perceived value and behavioural intention. Other researchers, Wen et al. (2005) and Jen and Hu (2003) also supported the findings. With that in mind, this study suspects positive relationship between perceived value and behavioural intention. Eleventh hypothesis of this study is stated as follows.

H11: Perceived value positively influences on paratransit passengers' behavioural intentions
Besides affecting satisfaction and behavioural intentions, perceived value is also shown to have a relationship with the trust. The studies by Ulaga and Eggert (2006) with the aim of exploring the relationship between value and relationship quality (such as commitment, satisfaction, and trust) proved that value has a positive relationship with trust. The other studies also confirmed these findings (Zhou 2008; Kim et al. 2008; Vonder der Heidt 2011, Sun and Chen 2012). Therefore, perceived value is thought to have positive effect on trust. Twelfth hypothesis of this study is stated as follows.

H12: Perceived value positively affects trust in the context of paratransit passengers

3.2.5 The Influence of Satisfaction on Behavioural Intention and Trust

In studies of consumer behaviour based on the Theory of Planned Behaviour (TPB), Lai and Chen (2009) explained that, in general, researchers propose major determinant of behavioural intentions is customer satisfaction. This explanation is also supported by some other experts like Crosby et al. (1990) which found that customer satisfaction will affect the decision to continue having relationships with service provider. In line with Crosby et al. Fornell (1992) also argued that satisfied customers tend to maintain their consumption patterns and will consume similar products or services.

Empirically, several studies have shown that customer satisfaction has a positive and direct relationship with behavioural intentions. For example, Molinari et al. (2008) who conducted a study on service providers found that increased customer satisfaction will have an impact on repurchase intention and positive word of mouth. In the context of public transport services, Wen et al. (2005) in his study proved that passenger's satisfaction positively affects passenger loyalty. Research conducted by Lai and Chen (2011) showed a direct and positive correlation between passenger satisfaction and behavioural intentions.

Besides affecting behavioural intention, customer satisfaction will also increase confidence, which is another form of trust (Morgan and Hunt 1994; Selnes 1998). More clearly, customer satisfaction has been identified as an important antecedent of trust (Ravald and Grönroos 1996; Selnes 1998).

Based on the previous explanation, satisfaction is suspected to affect paratransit passengers' trust and behavioural intention. Hypothesis thirteenth and fourteenth of this study are formulated as follows:

H13: Satisfaction positively affects paratransit passengers' trust

H14: Satisfaction positively affects paratransit passengers' behavioural intentions

3.2.6 The Influence of Trust on Behavioural Intention

Literature on relational marketing Relationship marketing literature has identified that trust is a key factor in maintaining customer relationships (Berry 1995; Palmatier et al. 2006; Aurier and N'goala 2010). There are some reseachers that attempted to examine the relationship between trust and behavioural intention and they found a positive relationship between the two (Liu et al. 2005; Lee 2005; Crosby et al. 1990; dalam Lee 2005; Ganesan 1994; Gefen 2000; Chiu et al. 2009). Similar findings were obtained by Wen et al. (2005) who conducted research in the context of public land transport services. Given this, the study assumes that trust has a positive influence on behavioural intention of paratransit passengers. Fifteenth hypothesis of this study is stated as follows:

H15: Trust positively affects paratransit passengers' behavioural intentions

References

Afzal, H., Khan, M. A., Rehman, K., Ali, I., & Wajahat, S. (2010). Consumer's trust in the brand: Can it be built through brand reputation, brand competence and brand predictability. *International Business Research, 3*(1), 43–51.

Ajzen, I. (1991). The theory of planned behavior. *Organizational Behavior and Human Decision Processes, 50*, 179–211.

Albanese, F. (2001). Merchandising and licensing to improve brand equity. The Coca-Cola case. *SYMPHONY A Emerging ucs in Management*, n. 1., N: 1593-0319, Retrived from http://dx.doi.org/10.4468/2001.1.06albanese

Alireza, F., Ali, K., & Aram, F. (2011). How quality, value, image, and satisfaction create loyalty at an Iran telecom. *International Journal of Business and Management, 6*(8), 271–279.

Alsheikh, L., & Bojei, J. (2012). Customer's perceived value to use mobile banking services. *International Conference on Management, Behavioral Sciences and Economics Issues (ICMBSE'2012),* Penang.

Alves, H., & Raposo, M. (2010). The influence of university image on student behaviour. *International Journal of Educational Management, 24*(1), 73–85.

Andreassen, T. W., & Lindestad, B. (1998). Customer loyalty and complex service: The impact of corporate image on quality, customer satisfaction and loyalty for customers with varying degree of service expertise. *International Journal of Service Industry Management, 9*, 7–23.

Armitage C. J., & Christian J. (2006). From attitudes to behavior: Basic and applied research on the theory of planned behaviour. In C. J Amirtage & J. Christian J. (Eds.), *Planned behaviour: The relationship between human thought and action.* New Jersey: Transaction Publishers.

Aurier, P., & N'Goala, G. (2010). The differing and mediating roles of trust and relationship commitment in service relationship maintenance and development. *Journal of the Academy of Marketing Science, 38*(3), 303–325.

Ball, D., Celho, P. S., & Machás, A. (2004). The role of communication and trust in explaining customer loyalty: An extension to the ECSI model. *European Journal of Marketing, 38*(9/10), 1272–1293.

Belch, G. E., & Belch, M. A. (2003). *Advertising and promotion: An integrated marketing communicationa perspective*, (6th ed.). New York: The McGraw-Hill Companies.

Berry, L. L. (1995). Relationship marketing of services—growing interest, emerging perspectives. *Journal of the Academy of Marketing Science, 23*(4), 236–245.

Bloemer, J., de Ruyter, Ko., & Peeters, P. (1998). Investigating drivers of bank loyalty: The complex relationship between image, service quality and satisfaction. *International Journal of Bank Marketing*, 16/7, 276–286.

Boulding, W., Kalra, A., Staelin, R., & Zeithaml, V. A. (1993). A dynamic process model of service quality: from expectations to behavioral intention. *Journal of Marketing Research, 30*, 7–27.

Brandão E. A., Muniz, R. M., Filho, C. G., Rocchiccioli, G. F., Souki, G. Q., & Livramento, R. (2011). Brand relationships on retailing: The impact of image on behavioral intentions of consumers. *Review Administration of UFSM, Santa Maria, 4*(1), 170–186.

Bruhn, M. (2003). *Relationship Marketing Management of Customer Relationship* UK: Prentice Hall. Pearson Education Limited Edinburgh Gate Harlow Exxex CM202JE England.

Butcher, K., Sparks, B., & O'Callaghan, F. (2001). Evaluative and relational infuences on service loyalty. *International Journal of Service Industry Management, 12*, 310–327.

Chen, Ching-Fu & Tsai, DungChun. (2007). How destination image & evaluative factors affect behavioral intentions? , Tourism Management, 28, 1115–1122

Chen, C. F., & Sai, M. H. (2008). Perceived value, satisfaction, and loyalty of TV travel product shopping: involvement as a moderator. *Tourism Management, 29*, 1166–1171.

Chien-Hsiung, L. (2011). A study on the relations between the brand image and customer satisfaction in catering businesses. *African Journal of Business Management*, *5*(18), 7732-7739, N 1993-8233 ©2011 Academic Journals, Retrieved from http://www.academicjournals.org/AJBM

Chiu, C. M., Chang, C. C., Cheng, H. L., & Fang, Y. H. (2009). Determinants of customer repurchase intention in online shopping. *Online Information Review, 33*(4), 761–784.

Choi, K. S., Cho, W. H., Lee, S., Lee, H., & Kim, C. (2004). The relationships among quality, value, satisfaction and behavioral intention in health care provider choice: A South Korean study. *Journal of Business Research, 57*, 913–921.

Chung, I-K., & Lee, M-M. (2003). A study of influencing factors for repurchase intention in internet shopping malls. *IEEE.* Print ISBN: 0-7695-1926-1.

Cox, P. (2010). *Moving people, sustainable transport development*. Cape Town: UCT Press.

Cronin, J., & Taylor, S. A. (1992). Measuring service quality: A reexamination and extension. *Journal of Marketing, 56*, 55–68.

Cronin, J. J., Brady, M. K., & Hult, G. T. M. (2000). Assessing the effects of quality, value, and customer satisfaction on consumer behavioral intentions in service environments. *Journal of Retailing, 76*(2), 193–218, 0022-4359.

Crosby, L.A., Evans, K., & Cowles, D. (1990). Relationship quality in services selling: An interpersonal influence perspective. *Journal of Marketing, 54*, 68–81.

Dodds, W. B. (1991). In search of value: How price and store name information influence buyer's product perceptions. *The Journal of Service Marketing, 5*(3), 27–36.

Earl, P. E. (2012). Experiential analysis of automotive consumption. *Journal of Business Research, 65*, 1067–1072.

Falk, T., Hammerschmidt, M., & Schepers, J. J. L. (2010). The service quality satisfaction link revisited: Exploring a symmetries and dynamics. *Journal of the Academy of Marketing Science, 38*(3), 288–302.

Flavián, C., Guinalíu, M. & Torres, E. (2005). The influence of corporate image on consumer trust: *A comparative analysis in traditional versus internet banking , Internet Research, 15* (4), 447–470.

Fornell, C. (1992). A national customer satisfaction barometer: The Swedish experience. *Journal of Marketing, 5*, 1–21.

Fornell, C., & Larcker, D. F. (1981). Evaluating structural equation models with unobservable variables and measurement error. *Journal of Marketing Research, 48*, 39–50.

Ganesan, S. (1994). Determinant of long-term orientation in buyer-seller relationships. *Journal of Marketing, 58*(2), 1–19.

Gefen, D. (2000). E-commerce: The role of familiarity and trust. *Omega, 28*(6), 725–737.

Gronroos, C. (1994). From marketing mix to relationship marketing: Towards a paradigm shift in marketing. *Management Decision, 32*(2), 4–20.

Grönroos, C. (1988). Service quality: The six criteria of good perceived service quality. *Review of Business, 9*(Winter), 10–13.

Ha, J., & Jang, S. (2010). Perceived values, satisfaction, and behavioral intentions: The role of familiarity in Korean restaurants. *International Journal of Hospitality Management, 29*, 2–13.

Hellier, P. K., Geursen, G. M., Carr, R. A., & Rickard, J. A. (2003). Costumer repurchase intention: A general structural equation model. *European Journal of Marketing, 16*, 1762–1800.

Hooley, G. J., & Cook, D. (1984). SIMS: A store image monitoring system. *International Journal of Advertising, 3*, 129–138.

Huang, Y-K. (2009). The effect of airline service quality on passengers' behavioural intentions using servqual scores: A Taiwan case study. *Journal of the Eastern Asia Society for Transportation Studies*, 8, 2330–2343.

Ismail, I., Haron, H., Ibrahim, D. N., & Isa, S. M. (2006). Service quality, client satisfaction and loyalty towards audit firms: Perceptions of Malaysian public listed companies. *Managerial Auditing Journal*, 21(7), 738–756.

Jen, W., & Hu, K. C. (2003). Application of perceived value model to identify factors affecting passengers' repurchases intention on city bus: A case of the Taipei metropolitan area. *Transportation, 30*, 307–327.

Jen, W., & Lu, T.-J. (2003). Effects of service quality, customer satisfaction and switching barriers on passenger behavioral intentions in scheduled coach service. *Proceedings of the Eastern Asia Society for Transportation Studies, 4*, 701–716.

Joewono, T. B., & Kubota, H. (2007). User satisfaction with paratransit in competition with motorization in Indonesia: Anticipation of future implications. *Transportation, 34*, 337–354.

Kanibir, H., & Nart, S. (2009). Investigating drivers of choice behavior: Corporate image, perceived risk and trust interactions through reputation management. *Journal of Global Strategic Management, 5*, 121–134.

Kassim, N., & Abdullah, N. A. (2010). The effect of perceived service quality dimensions on customer satisfaction, trust, and loyalty in e-commerce settings: A cross cultural analysis. *Asia Pacific Journal of Marketing and Logistics, 22*(3), 351–371.

Keaveney, S. M., Hernman, A., Befurt, R., dan Landwehr, J. R. (2012). The eyes have it: How a car's face influences consumer categorization and evaluation of product line extensions. *Psychology and Marketing, 29*(1), 36–51.

Keegan, W. J., & Green, M. C. (2011). *Global Marketing*. London: Prentice Hall. 6th Ed.

Kim, C., Zhao, W., & Yang, K. H. (2008). An empirical study on the integrated framework of e-crm in online shopping: Evaluating the relationships among perceived value, satisfaction, and trust based on customers' perspectives. *Journal of Electronic Commerce in Organizations, 6*(3), 1–19.

Kotler, Philip. (2003). *Marketing insights from A to Z : 80 concepts every manager needs to know*. New Jersey: Wiley.

Kuo, Y.-F., Wu, C.-M., & Deng, W.-J. (2009). The relationships among service quality, perceived value, customer satisfaction, and post-purchase intention in mobile value-added services. *Computers in Human Behavior, 25*, 887–896.

Kuruuzum, A., & Koksal, C. D. (2010). The impact of service quality on behavioral intention in hospitality industry. *International Journal of Business and Management Studies, 2*(1), 9–15.

Lai, W.-T., & Chen, C.-F. (2011). Behavioral intention of public transit passenger—the role of service quality, perceived value, satisfaction and involvement. *Transport Policy, 18*, 318–325.

Lee, T. M. (2005). The impact of perceptions of interactivity on customer trust and transaction intentions. *Journal of Electronic Commerce Research, 6*(3), 165–180.

Lien, T. B., & Yu, C. C. (2001). An integrated model for the effects of perceived product, perceived service quality, and perceived price fairness on customer satisfaction and loyalty. *Journal of Consumer Satisfaction, Disatisfaction, and Complaining Behaviour, 14*, 125–140.

Liou, Y., & Tsao, W-Y. (2010). A study on the service model of public transportation: Taiwan railways versus. Taiwan high speed rail corporation. *Service Science, 1*(2), 32–42.

Liu, C., Marchewka, J. T., Lu, J., & Yu, C-S. (2005). Beyond concern—a privacy-trust-behavioral intention. *Information and Management, 42*, 289–304.

Lovelock, C., et al. (2004). *Services marketing in Asia*. New Jersey: Prentice Hall.

Martín, S. S., & Camarero, C. (2009). How perceived risk affects online buying. *Online Information Review, 33*(4), 629–654.

Mayer, R. R., & Greenwood, E. (1980). *The Design of Social Policy Research*. New Jersey: Prentice-Hall.

McClure, S. M., Li, J., Tomlin, D., Cypert, K. S., Montague, L., & Montague, P. (2004). Neural correlates of behavioral preference for culturally familiar drinks. *Neuron, 44*, 379–387.

Meng, S. H., Liang, G. S., & Yang, S. H. (2011). The relationship of cruise image, perceived value, satisfaction, and post-purchase behavioral intention on Taiwanese tourists. *African Journal of Business Management, 5*(1), 19–29.

Milfelner, B., Snoj, B., & Korda, A. P. (2011). Measurement of perceived quality, perceived value, image, and satisfaction interrelations of hotel services: Comparison of tourists from Slovenia and Italy. *Druš. Istraž. Zagre God.* 20 (2011), BR. 3 (113), STR. 605–624.

Minkiewicz, J., Evans, J., Bridson, K., & Mavondo, F. (2011). Corporate image in the leisure services sector. *Journal of Service Marketing, 25*(3), 190–201.

Molinari, L. K., Abratt, R., & Dion, P. (2008). Satisfaction, quality and effects on repurchase and positive word-of-mouth behavioral intention in a B2B services context. *Journal of Services Marketing, 22*(5), 363–373.

Monroe, K. B., & Chapman, J. D. (1987). Framing effects on buyers' subjective product evaluations. *Advances in Consumer Research, 14*, 193–197.

Morgan, R. M., & Hunt, S. D. (1994). The commitment-trust theory of relationship marketing. *Journal of Marketing, 58*(3), 20–38. http://dx.doi.org/10.2307/1252308

Oh, H. (1999). Service quality, customer satisfaction, and customer value: A holistic perspective. *Hospitality Management, 18*, 67–82.

Ostrowski, P. L., O'Brien, T. V., & Gordon, G. L. (1993). Service quality and customer loyaty in the commercial airline industry. *Journal of Travel Research, 32*, 16–24.

Palacio, A. B., Meneses, G. D., & Perez, P. J. P. (2002). The configuration of the university image and its relationship with the satisfaction of students. *Journal of Educational Administration, 40*(5), 486–505.

Palmatier, R. W., Dant, R., Grewal, D., & Evans, K. R. (2006). Factors influencing the effectiveness of relationship marketing: A meta-analysis. *Journal of Marketing, 70*, 136–153.

Park, J. W., Robertson, R., & Wu, C. L. (2005). Investigating the effects of airline service quality on airline image and passengers' future behavioural intentions: Findings from Australian international air passengers. *The Journal of Tourism Studies, 16*(1), 2–11.

Petrick, J. F. (2004). The roles of quality, value, and satisfaction in predicting cruise passengers behavioural intention. *Journal of Travel Research, 42*, 397–407.

Petrick, J. F., & Backman, S. J. (2002). An examination of the construct of perceived value for the prediction of golf travellers' intention to revisit. *Journal of Travel Research, 41*, 38–45.

Ramzi, Al-R M, & Mohamed, B. (2010). Customer loyalty and the impacts of service quality: The case of five star hotels in Jordan. *International Journal of Human and Social Sciences, 5*(13), 886–892.

Randazzo, S. (2006). Subaru: The emotional myths behind the brand's growth. *Journal of Advertising Research*, 11–17.

Ravald, A., & Grönroos, C. (1996). The value concept and relationship marketing. *European Journal of Marketing, 30*(2), 9–30.

Ravichandran, K., Bhargavi, K., & Kumar, S. A. (2010). Influence of service quality on banking customers' behavioural intention. *International Journal of Economics and Finance*, 2(4), 18–28.

Ruiz, D. M., Gremler, D. D., Washburn, J. H., & Carrión, G. C. (2008). Service value revisited: specifying a higher-order, formative measure. *Journal of Business Research, 61*, 1278–1291.

Ryu, K., Han, H., & Kim, Tae-Hee. (2008). The relationships among overall quick-casual restaurant image, perceived value, customer satisfaction, & behavioral intentions . *International Journal of Hospitality Management, 27*, 459–469.

Ryu, K., Lee, H-R., & Kim, W. G. (2012). The influence of the quality of the physical environment, food, and service on restaurant image, customer perceived value, customer satisfaction, and behavioral intentions. *International Journal of Contemporary Hospitality Management*, 24(2), 200–223.

Sekaran, U., & Bougie, R. (2011). *Research Methods for Business: A Skill Building Approach*. UK: Wiley.

Selnes, F. (1998). Antecedents and consequences of trust and satisfaction in buyer-seller relationships. *European Journal of Marketing, 32*(3), 305–322.

Shafeiha, S., & S Saeednia, H. (2011). Studying the impact of customers' perceptions from service brand on customer value-loyalty process (Iran Insurance Company). 2010 *International Conference on Business and Economics Research*, 1, Kuala Lumpur: IACSIT Press, 124–128.

Sharma, P., Chen, I. S. N., & Luk, S. T. T. (2012). Gender and age as moderators in the service evaluation process. *Journal of Services Marketing, 26*(2), 102–114.

Sheu, T.-S. (2010). Exploring the differential affections of service quality, sacrifice, perceived value, and customer satisfaction on university students' favorable and unfavorable behavioral intention. *Journal of Quality, 17*(6), 483–500.

Shimp, T. A. (2010). *Integrated Marketing Communication in Advertising and Promotion* (8th ed.). Heinle: South-Western CENGAGE Learning.

Shukla, P. (2010). Effects of perceived sacrifice, quality, value, and satisfaction on behavioral intention in the service environment. *Services Marketing Quarterly, 31*, 466–484.

Solomon, M. R. (2012). *Consumer behavior: Buying, having and being* (10th ed.). London: Prentice Hall.

Stank, T. P., Goldsby, T. J., & Vickery, S. K. (1999). Effect of service supplier performance on satisfaction and loyalty of store managers in the fast food industry. *Journal of Operations Management, 17*, 429–447.

Sumaedi, S., Bakti, I. G. M. Y., & Yarmen, M. (2012). The empirical study of public transport passengers' behavioral intentions: the roles of service quality, perceived sacrifice, perceived value, and satisfaction (Case study: Paratransit passengers in Jakarta, Indonesia). *International Journal for Traffic and Transport Engineering, 2*(1), 83–97. UDC: 656.121.072(594).

Sun, S.-Y., & Chen, Y.-C. (2012). Constructing the reserach model of determinants of customer satisfaction in online shopping: from the perspectives of justice theory and value theory. *International Journal of Information Technology and Computer Science, 5*, 1–7.

Sureshchandar, G. S., Chandrasekharan, R., & Anantharaman, R. N. (2002). The relationship between service quality and customer satisfaction—a factor specific approach. *Journal of Services Marketing, 16*(4), 363–379.

Tuskej, U., Golob, U., & Podnar, K. (2013). The role of consumer-brand identification in building brand relationships. *Journal of Business Research, 66*(1), 53–59.

Udo, G. J., Bagchi, K. K., & Kirs, P. J. (2010). An assessment of customers' e-service quality perception, satisfaction and intention. *International Journal of Information Management, 30*, 481–492.

Ulaga, W., & Eggert, A. (2006). Relationship value and relationship quality: Broadening the nomological network of business-to-business relationships. *European Journal of Marketing, 40*(3), 311–327.

Von der Heidt, T. (2011). Modelling the complexity of e-loyalty: The role of e-value, e-trust, e-satisfaction, and e-commitment. *The (interim) 2011 ANZMAC Conference Proceedings.*

Wen, C H., Lan, L. W., & Cheng, H-L. (2005).Structural equation modelling to determine passenger loyalty toward intercity bus services. *Journal of the Transportation Research Board, 1927*, 249–255.

Wilcox, K., Roggenveen, A. L., & Grewal, D. (2011). Shall i tell you now or later? Assimilation and contrast in the evaluation of experiential products. *Journal of Consumer Research, 38*(4), 763–773.

Zeithaml, V. A. (1988). Consumer perceptions of price, quality and value: a means and model and synthesis of evidence. *Journal of Marketing, 52*, 2–22.

Zhou, T. (2008). The impact of perceived value on user acceptance of mobile commerce. *Proceedings of the 2008 International Symposium on Electronic Commerce and Security*, 237–240.

Zhou, Z., Zhang, Q., Su, C., & Zhou, N. (2012). How do brand communities generate brand relationships? Intermediate mechanisms. *Journal of Business Research, 65*, 890–895.

Zins, A. H. (2001). Relative attitudes and commitment in customer loyalty models—Some experiences in the commercial airline industry. *International Journal of Service Industry Management, 12*(3), 269–294.

Chapter 4
Research Methodology

Abstract This chapter explicates the research method used in this study. It contains the research object (paratransit in Jabodetabek), research population and sample, operational definition and scale of variables, pilot test, survey and data analysis method.

Keywords Methods · Population · Sample · Variable · Survey · Analysis

4.1 Research's Object: Paratransit in Jabodetabek

This research was conducted in order to understand paratransit passengers' behavioural intention in Jabodetabek. Thus, the object of the study is limited to the context of paratransit services held in Jabodetabek.

In Indonesia, paratransit is one of the city's public transportation mode recognized in the Act No. 22/2009 and Transportation Minister Decree KM. 35/2003. It stated that paratransit is a transportation mode from one place to another inside a single area or region that utilizes public bus or passenger's cars bounded for particular routes.

In Jabodetabek, paratransit consists of large, medium and small bus and common passenger cars. This study focused on paratransit with small bus and common passenger cars type. Small bus means a motor vehicle with a capacity of 9–16 passengers with normal size and distance between eats excluding the driver's seat with vehicle length range from 4 to 6.5 m while a common passenger car is any motor vehicle equipped with as much eight seats excluding the driver's seat, with or without luggage space (Minister of Transportation Decree KM. 35/2003).

The selection of paratransit with small bus and common passenger cars type as the focus of the study based on several considerations. First, the paratransit types serve twigs route, route linking the residential location with branch route, so users of this paratransit types are very likely also the users of other types of paratransit or even other public land transport modes, such as commuter train. Second, local

S. Sumaedi et al., *Public Transport Passengers' Behavioural Intentions*, SpringerBriefs in Business, DOI: 10.1007/978-981-4585-24-8_4,

government in Jabodetabek area is actively preparing a mass public transport policy, such as BRT and MRT. It is difficult for these types of public transport to reach twigs routes. Thus, in the future, the role of city transport buses and passenger cars remain important as feeder for mass transport. Third, paratransit is the kind of transportation mode that dominates public transportation in Jabodetabek, as given in Table 1.1.

4.2 Population and Sample

In accordance with the previous description, the study population is paratransit passengers in Jabodetabek. The population characteristics (gender, age, income, education or the frequency of use of public transportation) are unknown since the paratransit service provider and the associations of paratransit service provider did not collect and publish data. Given that, following previous researches on the same topic (e.g. Wen et al. 2005; Lai and Chen 2011), this study used convenience sampling method.

The sample size was determined based on the overall population in Jabodetabek because public land transport is intended for all levels of society. Central Bureau of Statistic (CBS) noted that the population of Jabodetabek in 2010 is approximately 15,921,884. Thus, to be able to represent the population of Jabodetabek, with alpha value of 5 and 4.1 % sampling error (www.surveysystem.com), the required sample size is 563. For the sake of simplicity, the sample size is then rounded up to 570 users of paratransit. The steps to obtain and ensure that the sample is paratransit users will be explained in Sect. 4.4.

The sample is the resident of Jabodetabek areas which are categorized as city and can represent the north, east, south, west and centre of Jabodetabek, namely the city of Jakarta, Bogor, Depok, Tangerang and Bekasi. The distribution of the sample is determined by proportionally divide and compare the number of residents in each city. Figure 4.1 shows the distribution.

4.3 Operationalization and Measures of Variables

The study includes seven variables, namely, image, perceived sacrifice, service quality, perceived value, satisfaction, trust and behavioural intention. It has been known that the seven variables are latent variable that cannot be measured directly, so that the observable variables (measurement indicators) are necessary to represent the variables (Sekaran and Bougie 2011).

The research variables are measured using multi-indicators approach, whereas the constructs developments based on the relevant literature (Churchill 1979; Nunnally 1978). All variables, except service quality, are measured as unidimensional scale. Based on a study conducted by Transportation Research Board

Table 4.1 Indicators of research variables

No	Variables	Indicators/Observed variables	References
1	Perceived sacrifice	(PS1) The paratransits waiting time is acceptable	Zeithaml (1988); Wen et al. (2005)
		(PS2) The paratransits travel time is acceptable	
		(PS3) Price of paratransits is cheap	
		(PS4) Easy to find paratransits	
2	Image	(IM1) A good impression of paratransit	Park et al. (2005)
		(IM2) A better image than its competitors	
		(IM3) A good image in the minds of passengers	
3	Perceived Service quality	*Safety (SF)*	Transportation Research Board (1999)
		(SQ1) Safety on board	
		(SQ2) Safety from crime while riding	
		(SQ3) Safety related to behaviour of other persons	Wen et al. (2005)
		Comfort (CM)	
		(SQ4) Comfort of the seats	
		(SQ5) Degree of noise and vibration on the paratransit	
		(SQ6) Degree of crowding on the paratransit	
		(SQ7) Comfortable temperatures on the paratransit	
		Performance and reliability (PR)	
		(SQ8) The paratransit engine is still powerful	
		(SQ9) The number of public transport is adequate	
		(SQ10) Wait time when transferring	
		(SQ11) Travel time by paratransit	
		(SQ12) The paratransit obedience to traffic regulations	
		(SQ13) The paratransit deliver to destination	
		Crews attitude (CA)	
		(SQ14) The driver/conductor are neat in appearance	
		(SQ15) The driver/conductor are willing to help passenger	
		(SQ16) The driver/conductor willing to respond to passenger request	

(continued)

Table 4.1 (continued)

No	Variables	Indicators/Observed variables	References
		(SQ17) The driver/conductor are understanding your needs when you make inquiries	
		(SQ18) The driver/conductor are courteous	
		(SQ19) The driver are skilled full	
		Condition of vehicles and facilities (CVF)	
		(SQ20) The paratransit has modern looking facilities and equipment	
		(SQ21) Cleanliness of the paratransit exterior	
		(SQ22) Cleanliness of the paratransit interior	
		(SQ23) The paratransit clean of graffiti	
4	Perceived value	(PV1) Compared with time and money has been given, the service was valuable	Wen et al. (2005)
		(PV2) At the price has been paid, the service was acceptable	
		(PV3) It was worth taking this bus carrier rather than the others	
5	Satisfaction	(SAT1) The perceived service was equivalent to ideal service	Wen et al. (2005)
		(SAT2) Delighted to take paratransits	
		(SAT3) Overall, I was satisfied with paratransits	
		(SAT4) The perceived service was than the expectation	
6	Trust	(TR1) Believe to the bus carrier provided the best service	Wen et al. (2005)
		(TR2) The bus carrier offered a constant quality of service	
		(TR3) Overall, trust to paratransits	
7	Behavioural intention	(BI1) Say positive things about paratransits to others people	Zeithaml et al. (1996)
		(BI2) Recommend paratransits to someone who seeks your advice	
		(BI3) Encourage friends and relative to do business with paratransits	
		(BI4) Consider paratransits your first choice to travel	
		(BI5) Do more business with paratransits in the next few years	

(1999) and Wen et al. (2005), service quality is treated as a multidimensional variable, which consists of five dimensions, namely safety, comfort, performance and reliability, crews' attitude and the condition of vehicles and facilities shows the indicators of each variable. More clearly, Table 4.1 gives all indicators of the research variables. In the measurement, all of the indicators are expressed by using positive sentences, including the perceived sacrifice. In addition, all of those constructs were measured using a seven-point Likert scale ranging from 'strongly disagree (1)' to 'strongly agree (7)'.

Perceived sacrifice shows a level of sacrifice made by the use of public transport. The use of positive wordings in this variable makes it differently interpreted. Higher value indicates lower perceived sacrifice. Consequently, it results on distinct interpretation of statistical coefficient between perceived sacrifice with other constructs. Furthermore, the relationship between perceived sacrifice and other variables (perceived value) is negative if the calculation results showed positive statistical correlation. Conversely, if the results of the statistical calculation are negative then the relationship between the two is interpreted as positive.

4.4 Survey

Data collection methods used in this study are a survey with questionnaires, whereas the respondents are paratransit passengers in Jabodetabek area. The questionnaire consists of three main sections. The first part is an introduction and instructions to fill out the questionnaire. The second section contains questions related to the demographic profiles. The third part is the questions related to assessment of research variables.

The sample is Jabodetabek residents who utilize paratransit. To ensure that, following previous studies (Wen et al. 2005; Lai and Chen 2011), the distribution of the questionnaires were carried out in the operation place of public transportations. In this regard, the questionnaires were distributed in city terminal or transit stops located in the north, east, south, west and centre of predetermined Jabodetabek (see Fig. 4.1).

4.5 Data Analysis Method

This study used analysis of Structural Equation Modelling (SEM) to test the hypotheses. 'SEM is a multivariate statistical approach that allows researchers to concurrently examine both the measurement and structural components of a model by testing the relationships among multiple independent and dependent constructs' (Geffen et al. 2000). As recommended by Hair et al. (2010), authors conducted two phase of analysis, the measurement model analysis and structural analysis model. The analysis was carried out with (software) LISREL 8.80.

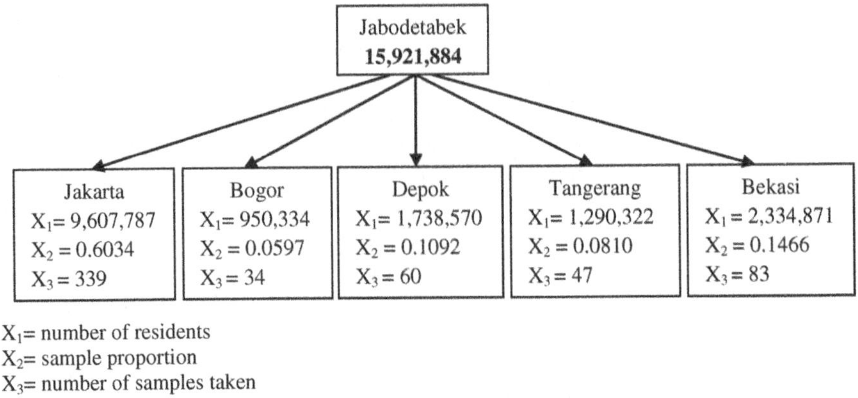

Fig. 4.1 Sample proportions and distributions

4.5.1 Measurement Model Analysis

Measurement model analysis was carried out in order to determine how well the model measures what is to be measured (Barett 2007). At this stage, three tests were performed, namely goodness of fit, validity and reliability. The method used was confirmatory factor analysis (CFA).

Goodness of fit was used to illustrate how well the model works as a set of united observations. There are many criteria options for assessing goodness of fit. In this research, the criteria used are the Root Mean Square Error of Approximation (RMSEA), Normed Fit Index (NFI), Non-Normed Fit Index (NNFI), Comparative Fit Index (CFI), Incremental Fit Index (IFI) and Relative Fit Index (RFI). The research model is said as fit (good) when the value of NFI, NNFI, CFI, IFI and RFI is greater than the cut value of 0.9 (Hair et al. 2010). As for the RMSEA criteria, the model is considered fit when RMSEA shows value of less than 0.8 (Hair et al. 2010).

The validity study utilized two measures, the convergent and discriminant validity. Convergent validity of the study variables is acceptable if (1) factor loading of each indicator ≥ 0.5 and significant at an alpha level of 5 % and (2) the results of composite reliability (CR) on any variable are above 0.6 (Hair et al. 2005). Instrument meets the discriminant validity if the AVE values are greater than the squared correlation between constructs (Fornell and Larcker 1981).

Reliability testing was performed by taking into account the value of Cronbach alpha (Cronbach 1951). The variable is considered reliable when Cronbach alpha value of each latent variable is greater than 0.6 (Hair et al. 2005).

4.5.2 Structural Model Analysis

'Structural models differ from measurement models in that emphasis moves from the relationships between latent constructs and measured variables to the nature and magnitude of the relationships between constructs (Hair et al. 2005)' (Fotopoulos and Psomas 2010). In other words, the purpose of the structural model testing is to test the research hypotheses conceptualized in this study. There are 15 hypotheses tested. The structural model tested is shown in Fig. 3.1.

Similar with measurement model analysis, structural analysis model also begins with the Goodness of fit testing. This test ensures that the structural model and the data used are fit for each other. Criteria and cut-off value of each criterion are no different from the ones used in measurement models analysis.

Another analysis performed in the structural model analysis is the analysis of the contribution of each factor to behavioural intention and analysis. Analysis of the contribution of each factor was done to see the direct, indirect and total effects of each factor on behavioural intention (Lai and Chen 2011).

References

Act No. 22 Year 2009 Concerning Road Traffic and Transport (In Indonesia language: Undang-Undang (UU) No. 22 Tahun 2009 tentang Lalu Lintas dan Angkutan Jalan).

Barrett, P. (2007). Structural equation modelling: Adjudging model fit. *Personality and Individual Differences, 4*, 815–824.

Churchill, G. A., Jr. (1979). A paradigm for developing better measures of marketing constructs. *Journal of Marketing Research, 16*, 64–73.

Cronbach, L. J. (1951). Coefficient alpha and the internal structure of test. *Psychometrica, 16*(3), 297–334.

Decree of the Minister of Transportation No KM 35 Year 2003 on the execution of public transportation (In Indonesia language: Keputusan Menteri Perhubungan No. KM. 35 Tahun 2003 tentang Penyelenggaraaan angkutan orang di jalan dengan kendaraan umum).

Fornell, C., & Larcker, D. F. (1981). Evaluating structural equation models with unobservable variables and measurement error. *Journal of Marketing Research, 48*, 39–50.

Fotopoulos, C. B., & Psomas, E. L. (2010). The structural relationships between TQM factors and organizational performance. *The TQM Journal, 22*, 539–552.

Gefen, D. (2000). E-commerce: The role of familiarity and trust. *Omega, 28*(6), 725–737.

Hair, J. F., Black, W. C., Babin, B. J., & Anderson, R. E. (2005). *Multivariate data analysis* (6th ed.). New Jersey: Prentice-Hall.

Hair, J. F., Black, W. C., Babin, B. J., & Anderson, R. E. (2010). *Multivariate data analysis* (7th ed.). New Jersey: Prentice-Hall.

Lai, Wen-Tai, & Chen, C.-F. (2011). Behavioral intention of public transit passenger—the role of service quality, perceived value, satisfaction and involvement. *Transport Policy, 18*, 318–325.

Nunnally, J. C. (1978). *Psychometric theory*. New York: McGraw-Hill.

Park, J. W., Robertson, R., & Wu, C. L. (2005). Investigating the effects of airline service quality on airline image and passengers' future behavioural intentions: Findings from australian international air passengers. *The Journal of Tourism Studies, 16*(1), 2–11.

Sekaran, U., & Bougie, R. (2011). *Research methods for business: A skill building approach*. UK: Wiley,

Transportation Research Board. (1999). A handbook for measuring customer satisfaction and service quality, *TRCP Report 47.*

Wen, C.-H., Lan, L. W., & Cheng, H.-L. (2005). Structural equation modelling to determine passenger loyalty toward intercity bus services. *Journal of the Transportation Research Board, 1927*, 249–255.

Zeithaml, V. A. (1988). Consumer perceptions of price, quality and value: A means and model and synthesis of evidence. *Journal of Marketing, 52*, 2–22.

Zeithaml, V. A., Berry, L. L., & Parasuraman, A. (1996). The behavioral consequences of service quality. *Journal of Marketing, 60*, 31–46.

Chapter 5
Results and Discussions

Abstract This chapter describes the research results. It comprises the result of confirmatory factor analysis and the upshot of structural model test.

Keywords Research result · Confirmatory factor analysis · Structural model test

5.1 Demographic Profile of Respondents

The number of respondents of this study is 570. The demographic profile can be seen in Table 5.1. Based on the table, it can be seen that 51 % of respondents were male. In terms of age, the majority of the respondents are under 20 years (40 %). The majority of respondents have a high school education background (50 %) and junior high school (25 %). The majority of respondents are students/college students (38 %). Furthermore, a majority of respondents (59 %) have an income of less than Rp 1,000,000 a month while only 5 % of respondents have income above Rp 5,000,000. In addition, the majority of respondents (42 %) use public transportations daily.

5.2 Respondents Assessment on The Study Variables

Descriptive statistics, including the average and standard deviation of each research variable indicators can be seen in Table 5.2. Regarding the indicators of behavioural intentions, respondents tend to be neutral in nature. It is indicated by their assessment towards behavioural intention (4.19–4.33) which fell between 4 and 4.5.

Based on the table, it can also be seen that the respondents' assessment related to indicators of perceived sacrifice ranged from 4.13 to 5.15. It reveals that respondents tend to have positive assessment of the perceived sacrifice (the value of 2 indicators are greater than 5). In addition, the assessment of indicators related

S. Sumaedi et al., *Public Transport Passengers' Behavioural Intentions*, SpringerBriefs in Business, DOI: 10.1007/978-981-4585-24-8_5,

Table 5.1 Demographic profiles

Sample characteristics	Percentage (%)
Sex	
• Male	51
• Female	49
Total	100
Age	
• ≤20 YO	40
• 21–30 YO	29
• 31–40 YO	18
• 41–50 YO	9
• ≥51 YO	3
Total	100
Last acquired education	
• SD (elementary school)	11
• SMP (junior high school)	25
• SLTA (high school)	50
• Perguruan tinggi (university)	15
Total	100
Occupation	
• Civil servant	2
• Entrepreneurs	15
• TNI/POLRI (military)	2
• Students	38
• Private sector	27
• Unemployed	16
Total	100
Monthly income	
• Les than Rp 1,000,000	59
• Rp 2,500,000–5,000,000	10
• Rp 1,000,000–2,500,000	28
• Above Rp 5,000,000	3
Total	100
Usage frequency	
• Daily	42
• Minimum of once a week	20
• 2–5 times a week	23
• Minimum of once a month	15
Total	100
Purpose	
• Going to work	26
• Shopping	33
• Going to school	5
• Social activities	26
• Other	11
Total	100

Table 5.2 Descriptive statistics

No	Variables	Indicators/observed variables	Mean	Std. Dev
1	Perceived sacrifice	(PS1) The paratransits waiting time was acceptable	4.13	1.598
		(PS2) The paratransits travel time was acceptable	4.09	1.465
		(PS3) Price of paratransits is cheap	5.15	1.583
		(PS4) Easy to find paratransits	5.15	1.466
2	Image	(IM1) A good impression of paratransit	4.12	1.430
		(IM2) A better image than its competitors	3.98	1.388
		(IM3) A good image in the minds of passengers	4.02	1.239
3	Perceived service quality	*Safety (SF)*		
		(SQ1) Safety on board	3.78	1.563
		(SQ2) Safety from crime while riding	3.41	1.646
		(SQ3) Safety related to behaviour of other persons	3.81	1.374
		Comfort (CM)		
		(SQ4) Comfort of the seats	3.57	1.479
		(SQ5) Degree of noise and vibration on the paratransit	3.57	1.478
		(SQ6) Degree of crowding on the paratransit	3.40	1.618
		(SQ7) Comfortable temperatures on the paratransit	2.92	1.479
		Performance & reliability (PR)		
		(SQ8) The paratransit engine is still powerful	4.02	1.432
		(SQ9) The number of public transport is adequate	4.86	1.607
		(SQ10) Wait time when transferring	4.32	1.511
		(SQ11) Travel time by paratransit	4.22	1.428
		(SQ12) The paratransit obedience to traffic regulations	3.40	1.662
		(SQ13) The paratransit deliver to destination	4.94	1.485
		Crews attitude (CA)		
		(SQ14) The driver/conductor are neat in appearance	3.34	1.463
		(SQ15) The driver/conductor are willing to help passenger	4.01	1.500
		(SQ16) The driver/conductor willing to respond to passenger request	4.19	1.475
		(SQ17) The driver/conductor are understanding your needs when you make inquiries	3.98	1.391
		(SQ18) The driver/conductor are courteous	3.89	1.461
		(SQ19) The driver are skilled full	4.27	1.586
		Condition of vehicles & facilities (CVF)		
		(SQ20) The paratransit has modern looking facilities and equipment	3.53	1.390
		(SQ21) Cleanliness of the paratransit exterior	3.78	1.474
		(SQ22) Cleanliness of the paratransit interior	3.91	1.402
		(SQ23) The paratransit clean of graffiti	3.32	1.645
4	Perceived value	(PV1) Compared with time and money has been given, the service was valuable	4.33	1.497

(continued)

Table 5.2 (continued)

No	Variables	Indicators/observed variables	Mean	Std. Dev
		(PV2) At the price has been paid, the service was acceptable	4.90	1.389
		(PV3) It was worth taking this bus carrier rather than the others	4.40	1.628
5	Satisfaction	(SAT1) The perceived service was equivalent to ideal service	3.82	1.612
		(SAT2) Delighted to take paratransits	4.35	1.440
		(SAT3) Overall, i was satisfied with paratransits	3.83	1.583
		(SAT4) The perceived service was than the expectation	3.26	1.440
6	Trust	(TR1) Believe to the bus carrier provided the best service	3.82	1.503
		(TR2) The bus carrier offered a constant quality of service	3.47	1.488
		(TR3) Overall, trust to paratransits	4.00	1.473
7	Behavioural intention	(BI1) Say positive things about paratransits to others people	4.19	1.360
		(BI2) Recommend paratransits to someone who seeks your advice	4.35	1.307
		(BI3) Encourage friends and relative to do business with paratransits	4.28	1.329
		(BI4) Consider paratransits your first choice to travel	4.33	1.329
		(BI5) Do more business with paratransits in the next few years	4.19	1.431

to image ranged from 3.98 to 4.04, indicating that respondents tend to judge paratransit's image negatively. The indicators of service quality are assessed variously by respondents (3.32–4.94). Respondents tended to give positive assessment on reliability and performance dimensions, in which 4 indicators are put from 4.02 to 4.94, while an indicator was rated 3.40. Respondents tend to slight positively assess perceived value indicators (4.33–4.90). On the other hand, respondents tended to judge satisfaction (3 indicators under 3.84 and only one indicator is assessed 4.35) and trust (2 indicators below 3.83 and 1 indicator was valued 4.00) indicators negatively.

Based on the above explanation, it can be seen that the respondents various range of assessments towards the indicators. Respondents gave positive assessment on the indicators of perceived sacrifice, perceived value and the dimensions of performance and reliability (service quality) while other indicators were given negative assessments. Respondents are neutral towards behavioural intention indicators.

Table 5.3 Goodness of fit (GF) index for measurement model

GF measures	Measurement value	Cut of value	Results GF	Reference
NFI	0.96	>0.09	Good	Hair et al. (2010)
NNFI	0.97	>0.09	Good	Hair et al. (2010)
CFI	0.97	>0.09	Good	Hair et al. (2010)
IFI	0.97	>0.09	Good	Hair et al. (2010)
RFI	0.96	>0.09	Good	Hair et al. (2010)
RMSEA	0.059	<0.1	Good	Hair et al. (2010)

5.3 The Results of Measurement Model Analysis

5.3.1 Goodness of Fit Index for Measurement Model

Table 5.3 shows the criteria of goodness of fit for measurement model. Based on the table, the value of RMSEA was 0.059 or below the cut-off value of 0.8 (Hair et al. 2010). Thus, the measurement model met the criteria of RMSEA. Furthermore, measurement model also met other goodness of fit criteria (NFI, NNFI, CFI, IFI and RFI) since the results of the test show values above the cut-off (0.09) (Hair et al. 2010). It can be concluded that the measurement model is fit with the data used.

5.3.2 Construct Validity

Table 5.4 shows the result of convergent validity for each latent and observed variable. Each construct met the criteria of convergent validity since (1) all observed variables have similar standardized factor loading or more than 0.5 and statistically significant and (2) the values of composite reliability (CR) are bigger than cut-off value of 0.6 (Hair et al. 2005).

Table 5.5 shows the results of discriminant validity test. The test shows that there are four constructs that met the criteria of discriminant validity, which are: image, service quality, trust and behavioural intention. Meanwhile, the other constructs, which are perceived value (the value of AVE 0.46), behavioral intention (0.43), are perceived sacrifice (0.41), did not because their AVE values are smaller than the squared correlation between constructs. Nevertheless, the constructs were still used because those indicators are theoretically accepted as the measures of the variables, including in the context of public transportation researches (Agung 2011).

Table 5.4 The results of convergent validity and reliability test

Latent constructs	Factor loading (*p* value)	Error variance	CA	CR	AVE
Perceive sacrifice (PS)			0.733	0.73	0.41
PS1	0.68 (17.04)	0.53			
PS2	0.66 (10.88)	0.56			
PS3	0.55 (13.13)	0.70			
PS4	0.66 (16.84)	0.56			
Image (IM)			0.867	0.88	0.70
IM1	0.76 (19.25)	0.43			
IM2	0.90 (28.79)	0.19			
IM3	0.85 (22.60)	0.28			
Service quality (SQ)			0.853	0.85	0.54
SQ1	0.72 (19.29)	0.48			
SQ2	0.73 (18.91)	0.46			
SQ3	0.76 (19.82)	0.42			
SQ4	0.71 (18.59)	0.50			
SQ5	0.75 (20.90)	0.43			
Perceived value (PV)			0.678	0.71	0.46
PV1	0.70 (16.84)	0.51			
PV2	0.77 (17.07)	0.41			
PV3	0.54 (12.07)	0.71			
Satisfaction (SAT)			0.835	0.84	0.56
SAT1	0.75 (23.06)	0.44			
SAT2	0.73 (21.79)	0.46			
SAT3	0.81 (25.43)	0.34			
SAT4	0.70 (19.01)	0.50			
Trust (TR)			0.840	0.84	0.64
TR1	0.81 (25.20)	0.34			
TR2	0.80 (21.88)	0.36			
TR3	0.78 (23.67)	0.39			
Behavioural intention (BI)			0.775	0.78	0.43
BI1	0.64 (15.51)	0.59			
BI2	0.79 (20.49)	0.38			
BI3	0.74 (18.31)	0.46			
BI4	0.56 (11.61)	0.69			
BI5	0.50 (10.70)	0.75			

Table 5.5 The results of discriminant validity test

Latent constructs	AVE	PS	IM	SQ	PV	SAT	TR	BI
Perceive sacrifice (PS)	0.41	0.41						
Image (IM)	0.70	0.18	0.70					
Service quality (SQ)	0.54	0.23	0.38	0.54				
Perceived value (PV)	0.46	0.50	0.29	0.22	0.46			
Satisfaction (SAT)	0.56	0.40	0.30	0.34	0.36	0.56		
Tust (TR)	0.64	0.35	0.44	0.35	0.38	0.62	0.64	
Behavioral intention (BI)	0.43	0.26	0.35	0.26	0.31	0.26	0.31	0.43

Table 5.6 Goodness of fit index for structural model

GF measures	Measurement value	Cut of value	Results GF	Reference
NFI	0.96	>0.09	Good	Hair et al. (2010)
NNFI	0.97	>0.09	Good	Hair et al. (2010)
CFI	0.97	>0.09	Good	Hair et al. (2010)
IFI	0.97	>0.09	Good	Hair et al. (2010)
RFI	0.96	>0.09	Good	Hair et al. (2010)
RMSEA	0.06	<0.1	Good	Hair et al. (2010)

5.3.3 *Reliability of Constructs*

Based on the results shown in Table 5.4, each construct in this study also met the criteria of reliability since the Cronbach alpha (CA) value at each latent variable has a value greater than 0.6 (Hair et al. 2005). Based on the descriptions contained in Sect. 5.3, it can be concluded that the measurement model used in this study is fit with the data used, reliable and consistent.

5.4 Structural Model Analysis

The main objective of structural model analysis is to see if a research hypothesis is statistically acceptable or not (Anderson and Gerbing 1988). This is done by putting the value of t test (t-value) as the subject of significance test. Hypothesis is accepted if the number of absolute value of $t > 1.96$ ($alpha = 0.05$) with theoretically correct coefficient signs (positive or negative). Nevertheless, it should be confirmed beforehand that the structural model is fit with the data used.

5.4.1 *Goodness of Fit Index for Structural Model*

The value of each goodness of fit criterion for the structural model can be seen in Table 5.6. From the table we can see that this study has appropriate structural model because the values of all goodness of fit criteria exceeded the cut-off value (Hair et al. 2010).

5.4.2 *Results of Hypotheses Testing*

The results of the structural model analysis can be seen in Fig. 5.1. Based on the figure, 9 hypotheses are proven to be statistically significant while the rest (H3, H7, H9, H12, H14 and H15) are not. Table 5.7 shows the results of hypotheses testing in details.

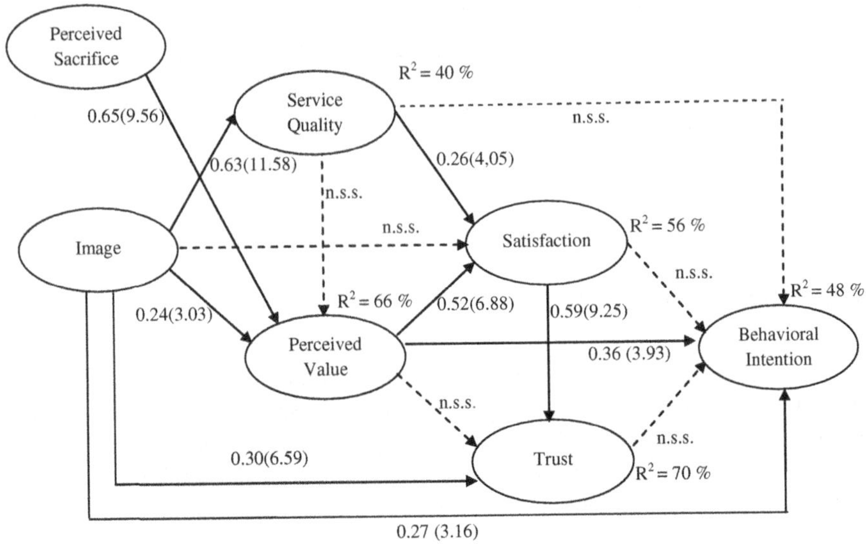

Fig. 5.1 Results of structural model analysis. *Note* n.s.s. = non satistically significant

Table 5.7 Results of significance test

Hypotheses	Influences of latent variables	t-value	Coefficients	Significance
H1	Perceived sacrifice → Perceived value	9.56	0.65	Accepted
H2	Image → Service quality	11.58	0.63	Accepted
H3	Image → Satisfaction	1.22	0.11	Rejected
H4	Image → Perceived value	3.03	0.24	Accepted
H5	Image → Behavioural intention	3.16	0.27	Accepted
H6	Image → Trust	6.59	0.30	Accepted
H7	Service quality → Perceived value	1.25	0.07	Rejected
H8	Service quality → Satisfaction	4.05	0.26	Accepted
H9	Service quality → Behavioural intention	1.80	0.12	Rejected
H10	Perceived value → Satisfaction	6.88	0.52	Accepted
H11	Perceived value → Behavioural intention	3.93	0.36	Accepted
H12	Perceived value → Trust	1.05	0.06	Rejected
H13	Satisfaction → Trust	9.25	0.59	Accepted
H14	Satisfaction → Behavioural intention	−0.39	−0.04	Rejected
H15	Trust → Behavioural intention	1.04	0.11	Rejected

The first hypothesis of this study stated that perceived sacrifice negatively affects perceived value. The results show that the hypothesis is accepted. In the context of public land transport service, this finding is consistent with the results from studies conducted by Lai and Chen (2011) and Wen et al. (2005). Thus, this study confirms that higher perceived sacrifices results in lower perceived value of paratransit service. Conversely, lower perceived sacrifices leads to higher perceived value of paratransit service.

The second hypothesis of this study stated that image has positive impact on perceived value. In other words, this study assumed that favourable image will result in higher perceived value. On the other hand, unfavourable image leads to lower perceived value. The results confirm the hypothesis. Thus, this finding supports the view of Andreassen and Lindestad (1998).

The third hypothesis of this study stated that image has positive impact on paratransit passengers' satisfaction. The results show that the third hypothesis is not supported. Thus, favourable image on paratransit services is not proven to improve passengers' satisfaction. According to Andreassen and Lindestad (1998), image is formed in consumers mind through communication and experience. Formation of image leads to expectations. In conjunction with these findings, passengers' satisfactions are not affected by image because passengers have different expectations of the paratransit services. Thus, different level of expectations results in various level of satisfaction. This is consistent with the findings of Chen and Tsai (2007) which stated that image cannot directly affect customer satisfaction, but, image can affect customer satisfaction through the mediating variable such as service quality and perceived value.

The fourth hypothesis of this study stated that image has a positive effect on perceived value of paratransit passengers. In other words, this study assumed that positive image of paratransit services can increase passengers' perceived value. Inversely, poor image of paratransit can decrease the perceived value. Based on the analysis, this research found that the fourth hypothesis can be accepted. The fourth hypothesis supported previous research by Milfelner et al. (2011) by Ryu et al. (2012), Shafeiha and Saeednia (2011) and Alireza et al. (2011). Furthermore, Meng et al. (2011), Liou and Tsao (2010) and Zins (2001) also support this finding to in context of public transport services.

The fifth hypothesis of this study stated that image has positive influence on paratransit passengers' behavioural intentions. The authors suspected that the change in passengers' behavioural intentions is significantly influenced by the image of paratransit service. Based on the analysis, the authors found that the fifth hypothesis of this study is accepted. Thus, the finding of this study has reinforced research conducted before by Brandão et al. (Brandão et al. 2011), Chen and Tsai (2007) and Ryu et al. (2008). Furthermore, Meng et al. (2011) also supported this finding in the context of public transport services.

The sixth hypothesis proposed image as a positive influence on paratransit passengers' trust. The results showed that the sixth hypothesis of this study is accepted. Thus, a positive image of paratransit will increase passengers' trust. On the other hand, poor image of paratransit services can reduce passengers' trust. The sixth hypothesis of this study is in line with findings from researches conducted by Flavian et al. (2005) and Ball et al. (2004).

Seventh hypothesis proposed that service quality has positive effect on perceived value. Based on the analysis, this hypothesis is rejected. The assumption that increasing the quality of paratransit services will increase perceived value is not statistically proven. In other words, in the context of perceived value, passengers apparently are not so keen on improved service quality. From these results can also be interpreted also that service value sought by passengers is not on the aspect service quality, but in other aspects such as perceived sacrifice. Perceived value is the comparison between benefits and sacrifice of paratransit. On this term, consumers can see the value of the service from three points of view (1) benefits are more important than the sacrifice, (2) sacrifice are more important than the benefits (3) both are equally important. Related to this research, passengers might see that sacrifice is more important than the benefits. This allegation can be hold true since most of the study's respondents came from lower-middle class. The rejection of this finding is also supported by Liou and Tsao (2010).

Eighth hypothesis stated that service quality positively affects paratransit passengers' satisfaction. The authors suspected that the increase in service quality has a significant impact perceived value. The results show that the eighth hypothesis of this study is accepted. The result is in line with previous findings, as Cronin and Taylor (1992), Ostrowski et al. (1993), Stank et al. (1999), Butcher et al. (2001) and Hellier et al. (2003). In fact, some researchers suggested that this finding can be applied in the context of public transport services (e.g. Wen et al. 2005; Joewono and Kubota 2007; Lai and Chen 2011).

The statement that service quality positively affects paratransit passengers' behavioural intentions is the ninth hypothesis. The hypothesis is not supported. In that case, in the context of paratransit service, changes in service quality did not provide significant impact on passengers' behavioural intentions. The correlation between the two constructs is still being debated among researchers, as some researchers have found that service quality has a direct effect on behavioural intentions (Lien and Yu 2001; Al-Rousan et al. 2010; Lai and Chen 2011; Rahnama et al. 2012; Malik et al. 2011; Ramzi and Mohamed 2010) while other researchers said otherwise (Chang and Wang 2011; Cheng et al. 2008; Liao 2012; Kiran and Diljit, 2011; Alireza et al. 2011; Chumpitaz and Paparoidamis 2004; Bei and Chiao 2001; Cronin and Taylor 1992). Related to the results of this study, the finding of the authors support the lack of correlation between service quality and paratransit passengers' behavioural intentions.

Perceived value positively affects satisfaction is the tenth hypothesis proposed in this study. From the results of the analysis found that the change in satisfaction is significantly affected by change in perceived value. Thus, this study accepted the tenth hypothesis in the context of paratransit service. The acceptance of the tenth hypothesis of this study support previous findings (Andreassen and Lindestad 1998 and Hellier et al. 2003) even in the context of paratransit service (Lai and Chen 2011; Wen et al. 2005).

Eleventh hypothesis of this study stated that perceived value positively influences paratransit passengers' behavioural intentions. In this hypothesis, the researchers alleged that increased passengers' behavioural intentions significantly affected by increased perceived value. From the analysis, authors found that the eleventh hypothesis of this study is accepted. This supported the findings of Petrick and Backman (2002), Petrick (2004), Chen and Tsai (2007) and Meng et al. (2011). The other researchers, such as Lai and Chen (2011), Wen et al. (2005) and Jen and Hu (2003) also supported these results in the context of public land transport service.

In this study, the authors proposed the twelfth hypothesis as 'perceived value positively affects the trust of paratransit passengers'. This hypothesis is rejected. In that case, the results of this study indicate that changes in perceived value passengers did not significantly affect paratransit passengers' trust. In other words, the trust sought by passenger is not on the aspect of perceived value of public transportation services. Trust desired might be skewed on the aspect of security and safety.

The thirteenth hypothesis of this study stated that satisfaction positively affects paratransit passengers' trust. This study assumed that the change in satisfaction gives a significantly positive impact on the change in passengers' trust. From the analysis it is found that the thirteenth hypothesis is supported by the data. In that case, the research is consistent with several previous studies (Molinari et al., 2008; Caceres and Paparoidamis 2007; Selnes 1998; Yeh and Li 2009; Sahadev and Purani 2008).

Fourteenth hypothesis of this study stated that satisfaction positively affects paratransit passenger's behavioural intentions. The fourteenth hypothesis of this study is not supported. Thus, the finding of this study suggested that increased satisfaction is not significantly affected paratransit passengers' behavioural intentions and consequently, decreased satisfaction also fails to affect paratransit passengers' behavioural intentions. Kang and Wang (2009), Sahadev and Purani (2008) and Andreassen and Lindestad (1998) also found the same finding. Furthermore, the result implied that increased passengers' satisfaction does not guarantee repurchase and intention to recommend.

Last hypothesis of this study stated trust positively affects paratransit passengers' behavioural intentions. Researchers suggested that the increased passengers' behavioural intention is significantly affected by the increase in trust. From the analysis, the authors found that the hypothesis of the fifteenth study is not supported. The finding also corresponds with previous studies by Haghighi et al. (2012), Hashim et al. (2011) and Eid (2011).

Referring to the hypotheses test results, it can be concluded that the factors that directly influence behavioural intentions are perceived value and image. Further it is also found that image and perceived sacrifice has indirect influence on behavioural intention. Meanwhile, other factors such as service quality, satisfaction and trust are not shown to affect behavioural intention, either directly or indirectly. In other words, the model of behavioural intention confirmed by this research is a model that shows that behavioural intention is influenced directly by image and perceived value and indirectly affected by perceived sacrifice and image through perceived value.

Table 5.8 Direct, indirect and total effect of behavioural intentions

Path	Effect	Estimates	T-value
Perceived sacrifice → Behavioural intentions	Direct effect	–	–
	Indirect effect	0.24	5.37
	Total effect	0.24	5.37
Image → Behavioural intentions	Direct effect	0.27	3.16
	Indirect effect	0.23	3.60
	Total effect	0.50	8.01
Perceived value → Behavioural intentions	Direct effect	0.36	3.93
	Indirect effect	0.01	0.25
	Total effect	0.37	5.24

5.4.3 Contribution of Each Factor on Behavioural Intentions

To determine the contribution of each factor, we sum of the direct and indirect effects (Lai and Chen 2011). In accordance with the discussion in previous subsection, there are three factors related to behavioural intention of paratransit passengers, perceived sacrifice, perceived value and image. The contributions of these factors to behavioural intention can be seen in Table 5.8.

From Table 5.8 we can see that the image has the greatest total effect on behavioural intention (0.50), followed by perceived value (0.36) and perceived sacrifice (0.24). This condition indicates that image, which represents hedonics variables, has a very important role in the formation of behavioural intention. In the direct effect category, perceived value (0.36) has the biggest effect on behavioural intentions while image (0.27) is in the second position. From the category of indirect effect, perceived sacrifice (0.24) has the greatest indirect effect on behavioural intention and image (0.23) is still in the second place.

5.4.4 Coefficient of Determination (R^2)

Image, perceived sacrifice and perceived value are important factors that influence paratransit passengers' behavioural intention. Based on analysis results, the value of coefficient of determination (R^2) for behavioural intention is 48 %. It means that the influences of image, perceived sacrifice and perceived value can account for 48 % of passengers' behavioural intention and the rest might be explicable using other factors not considered in this study. Figure 5.1 shows the value of coefficient of determination for each endogen variables, namely trust (70 %), satisfaction (56 %), perceived value (66 %) and service quality (40 %).

References

Agung, I. G. N. (2011). *Cross section & experimental data analysis using eviews* (1st ed.). Wiley.

Alireza, F., Ali, K., & Aram, F. (2011). How quality, value, image, and satisfaction create loyalty at an iran telecom. *International Journal of Business & Management, 6*(8), 271–279.

Al-Rousan, Ramzi, M., & Mohamed, B. (2010). Customer loyalty & the impacts of service quality: the case of five star hotels in Jordan. *International Journal of Human & Social Sciences, 5*(13), 886–892.

Andreassen, T. W., & Lindestad, B. (1998). Customer loyalty and complex service: The impact of corporate image on quality, customer satisfaction and loyalty for customers with varying degree of service expertise. *International Journal of Service Industry Management, 9*, 7–23.

Anderson, J. C., & Gerbing, D. W. (1988). Structural equation modeling in practice: A review & recommended two-step approach. *Psychological Bulletin, 103*(3), 411–423.

Ball, D., Celho, P. S., & Machás, A. (2004). The role of communication & trust in explaining customer loyalty: An extension to the ECSI model. *European Journal of Marketing, 38*(9/10), 1272–1293.

Bei, L.-T., & Yu-Ching, C. (2001). An integrated model for the effects of perceived product, perceived service quality, & perceived price fairness on consumer satisfaction & loyalty. *Journal of Consumer Satisfaction, Dissatisfaction & Complaining Behaviour, 14*, 125–139.

Brandão, E. A., Muniz, R. M., Filho, C. G., Rocchiccioli, G. F., Souki, G. Q., & Livramento, R. (2011). Brand relationships on retailing: The impact of image on behavioral intentions of consumers. *Revista de Administração. UFSM, Santa Maria, 4* (1), 170–186.

Butcher, K., Sparks, B., & O'Callaghan, F. (2001). Evaluative & relational influences on service loyalty. *International Journal of Service Industry Management, 12*, 310–327.

Caceres, R. C., & Paparoidamis, N. G. (2007). Service quality, relationship satisfaction, trust, commitment & business-to-business loyalty. *European Journal of Marketing, 41*(7), 836–867.

Chang, H. H., & Wang, H.-W. (2011). The moderating effect of customer perceived value on online shopping behaviour. *Online Information Review, 35*(3), 333–359.

Chen, C.-F., & Tsai, D. C. (2007). How destination image & evaluative factors affect behavioral intentions? *Tourism Management, 28*, 1115–1122.

Cheng, T. C. E., Lai, I. C. F., & Young, A. C. L. (2008). The driving forces of customer loyalty: A study of internet service providers in Hong Kong. *International Journal of E-Business Research, 4*(4), 26–42.

Chumpitaz, R., & Paparoidamis, N. G. (2004). Service quality & marketing performance in business-to-business markets: exploring the mediating role of client satisfaction. *Managing Service Quality, 14*(2/3), 235–248.

Cronin, J. J., & Taylor, S. A. (1992). Measuring service quality: a reexamination & extension. *Journal of Marketing, 56*, 55–68.

Eid, M. I. (2011). Determinants of e-commerce customer satisfaction, trust, & loyalty in Saudi Arabia. *Journal of Electronic Commerce Research, 12*(1), 78–93.

Flavian, C., Guinalíu, M., & Torres, E. (2005). The influence of corporate image on consumer trust: A comparative analysis in traditional versus internet banking. *Internet Research, 15*(4), 447–470.

Haghighi, M., Dorosti, A., Rahnama, A., & Hoseinpour, A. (2012). Evaluation of factors affecting customer loyalty in the restaurant industry. *African Journal of Business Management, 6*(14), 5039–5046.

Hair, J. F., Black, W. C., Babin, B. J., & Anderson, R. E. (2005). *Multivariate data analysis* (6th ed.). New Jersey: Prentice Hall.

Hair, J. F., Black, W. C., Babin, B. J., & Anderson, R. E. (2010). *Multivariate data analysis* (7th ed.). New Jersey: Prentice-Hall.

Hashim, R. A., Mashudi, P. M., Jahit, Z. H. H., Rashid, S., Abu, N. K., & Ahmad, R. H. J. (2011). Influence of relationship quality on hotel guests' loyalty: A case study of a malaysian budget hotel. *Canadian Journal on Scientific & Industrial Research, 2*(6), 220–229.

Hellier, P. K., Geursen, G. M., Carr, R. A., & Rickard, J. A. (2003). Costumer repurchase intention: A general structural equation model. *European Journal of Marketing, 16*, 1762–1800.

Jen, W., & Hu, K. C. (2003). Application of perceived value model to identify factors affecting passengers' repurchases intention on city bus: A case of the Taipei metropolitan area. *Transportation, 30*, 307–327.

Joewono, T. B., & Kubota, H. (2007). User satisfaction with paratransit in competition with motorization in Indonesia: anticipation of future implications. *Transportation, 34*, 337–354.

Kang, H.-C., & Wang, Y.-W. (2009). The relationships among sea-food restaurant service quality, perceived value, customer satisfaction & behavioral intentions using structural equation models: A case of pengh. *The 9th International Conference on Electronic Business* (629–933).

Kiran, K., & Diljit, S. (2011). Antecedents of customer loyalty : Does service quality suffice. *Malaysian Journal of Library & Information Science, 16*(2), 95–113.

Lai, W.-T., & Chen, C.-F. (2011). Behavioral intention of public transit passenger—the role of service quality, perceived value, satisfaction & involvement. *Transport Policy, 18*, 318–325.

Liao, K.-H. (2012). Service quality, & customer satisfaction: direct & indirect effects in a b2b customer loyalty framework. *The Journal of Global Business Management, 8*(1), 86–93.

Lien, T. B., & Yu, C. C. (2001). An integrated model for the effects of perceived product, perceived service quality, & perceived price fairness on customer satisfaction & loyalty. *Journal of Consumer Satisfaction, Dissatisfaction, & Complaining Behaviour., 14*, 125–140.

Liou, Y., & Tsao, W.-Y. (2010). A study on the service model of public transportation: Taiwan railways vs Taiwan high speed rail corporation. *Service Science, 1*(2), 32–42.

Malik, M. E., Naeem, B., & Arif, Z. (2011). Impact of perceived service quality on banking customers' loyalty. *Interdisciplinary Journal of Contemporary Research in Business, 3*(8), 637–645.

Meng, S. H., Liang, G. S., & Yang, S. H. (2011). The realationship of cruise image, perceived value, satisfaction, & post-purchase behavioral intention on Taiwanese tourists. *African Journal of Business Management, 5*(1), 19–29.

Milfelner, B., Snoj, B., & Korda, A. P. (2011). Measurement of perceived quality, perceived value, image, & satisfaction interrelations of hotel services: Comparison of tourists from Slovenia & Italy, *DRUŠ. ISTRAŽ. ZAGRE GOD.* 20 (2011). BR. 3 (113). STR. 605–624.

Molinari, L. K., Abratt, R., & Dion, P. (2008). Satisfaction, quality & effects on repurchase & positive word-of-mouth behavioral intention in a B2B services context. *Journal of Services Marketing, 22*(5), 363–373.

Ostrowski, P. L., O'Brien, T. V., & Gordon, G. L. (1993). Service quality & customer loyaty in the commercial airline industry. *Journal of Travel Research, 32*, 16–24.

Petrick, J. F., & Backman, S. J. (2002). An examination of the construct of perceived value for the prediction of golf travellers' intention to revisit. *Journal of Travel Research, 41*, 38–45.

Petrick, J. F. (2004). The roles of quality, value, & satisfaction in predicting cruise passengers behavioural intention. *Journal of Travel Research, 42*, 397–407.

Rahnama, A., Alaei, A., Shafaee, J., & Hamdam, H. (2012). Evaluating the impact of banking services quality on customer loyalty in mellat bank ardebil province. *Journal of Basic and Applied Scientific Research, 2*(3), 2498–2506.

Ryu, K., Han, H., & Kim, T.-H. (2008). The relationships among overall quick-casual restaurant image, perceived value, customer satisfaction, & behavioral intentions. *International Journal of Hospitality Management, 27*, 459–469.

Ryu, K., Lee, H.-R., & Kim, W. G. (2012). The influence of the quality of the physical environment, food, & service on restaurant image, customer perceived value, customer satisfaction, & behavioral intentions. *International Journal of Contemporary Hospitality Management, 24*(2), 200–223.

Sahadev, S., & Purani, K. (2008). Modelling the consequences of e-service quality. *Marketing Intelligence & Planning, 26*(6), 605–620.

Selnes, F. (1998). Antecedents and consequences of trust and satisfaction in buyer-seller relationships. *European Journal of Marketing, 32*(3), 305–322.

Shafeiha, S. & Saeednia, H. (2011). Studying the impact of customers' perceptions from service brand on customer value-loyalty process (Iran Insurance Company), 2010 *International Conference on Business & Economics Research* 1, (pp. 124–128) Kuala Lumpur: IACSIT Press.

Stank, T. P., Goldsby, T. J., & Vickery, S. K. (1999). Efect of service supplier performance on satisfaction & loyalty of store managers in the fast food industry. *Journal of Operations Management, 17*, 429–447.

Wen, C.-H., Lan, L. W., & Cheng, H.-L. (2005). Structural equation modelling to determine passenger loyalty toward intercity bus services. *Journal of the Transportation Research Board, 1927*, 249–255.

Yeh, Y. S., & Li, Y.-M. (2009). Building trust in m-commerce: Contributions from quality & satisfaction. *Online Information Review, 33*(6), 1066–1086.

Zins, A. H. (2001). Relative attitudes & commitment in customer loyalty models–some experiences in the commercial airline industry. *International Journal of Service Industry Management, 12*(3), 269–294.

Chapter 6
Research Implications

Abstract This chapter relates the implications of this research. The implications conferred in the chapter are theoretical, managerial and policy implications.

Keywords Theoretical implications · Managerial implications · Policy implications

6.1 Theoretical Implications

The research was carried out in order to understand passengers' behavioural intention through Structural Equation Modelling (SEM). From various hypotheses proposed by this study, it was found that paratransit passengers' behavioural intentions are affected only by perceived sacrifice, image, and perceived value. The relationships between those four variables are shown in Fig. 6.1.

Figure 6.1 shows that paratransit passengers' behavioural intention is directly affected by image, whereas the influence is positive. In addition, image also indirectly influence behavioural intention through perceived value, whereas the influence is positive. Perceived value is also negatively affected by perceived sacrifice. On other words, perceived sacrifice influences behavioural intention indirectly through perceived value.

This study also calculated the contribution of perceived sacrifice, perceived value, and image on behavioural intention and the results can be seen in Table 5.8. From Table 5.8, we can see that the total effect of image on behavioural intention is larger than the other two factors.

Referring to the above explanation, it can be stated that the model of behavioural intention found in this study emphasizes the importance of image in shaping passengers' behavioural intention. Image is a variable that represents the performance of service providers in meeting hedonic experiential needs of its customers (Falk et al. 2010; Solomon 2012). Therefore, we refer the behavioural intention model of this research as behavioural intention model based on hedonic experiential needs or hedonic model in short.

S. Sumaedi et al., *Public Transport Passengers' Behavioural Intentions*, SpringerBriefs in Business, DOI: 10.1007/978-981-4585-24-8_6,

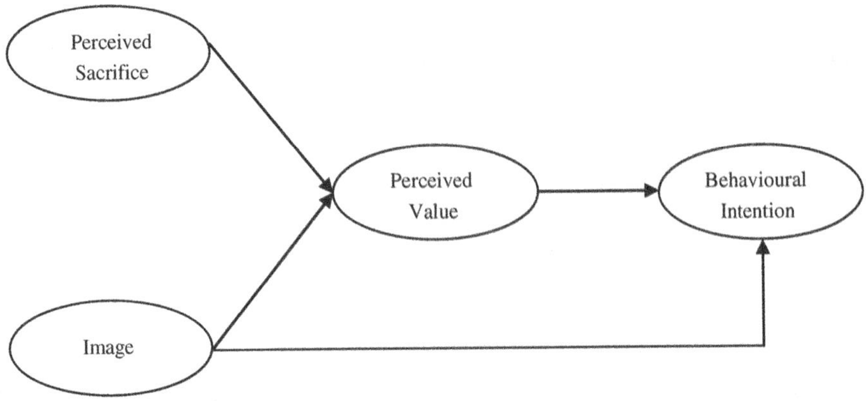

Fig. 6.1 Behavioural Intention Model of Paratransit Passengers in Jabodetabek

Based on the Hedonic Model, we can explain the phenomenon of a paratransit passengers' behavioural intention as follows. A passengers will have favourable behavioural intention towards paratransit service if she or he feels that the service is both valuable and has positive image. Conversely, paratransit passengers with unfavourable perceived value and negative image on the service will have unfavourable behavioural intention. In addition, this can explain satisfactorily why the descriptive statistic (Table 5.2) shows that paratransit passengers in Jabodetabek have a neutral behavioural intention (the assessment value ranges from 4.19 to 4.33), which are the negative image on paratransit and just 'slight' positive perceived value.

Perceived value and image has complementary role in forming favourable behavioural intention. On one hand, positive perceived value will make people interested in using paratransit service. On the other hand, positive image of paratransit service will negate switching intention or 'lock' the passenger to switch to other alternatives. Further, image has bigger role in fostering passengers' behavioural intention than perceived value.

Hedonic model also indicates that a person will see the value of paratransit service based on the comparison between the image of the service and the sacrifices he or she must do in order to get the service. A passenger will perceive the value of such service positively if he or she feels the image of the service exceeds the sacrifices. One of the unique findings of this research is the nonexistent role of service quality as the determinant of perceived value. It might be due to the fact that in the context of hedonic experiential needs, assessment of service quality tends to be biased because of the influence of image (Shimp 2010; Solomon 2012; Wilcox et al. 2011).

Nevertheless, it does not mean that service quality should be neglected because at some point, even people with hedonic experiential needs still care about minimum functional aspect (Falk et al. 2010) which is the core benefit of product (see Kotler and Keller (2012) in discussion about product classification). In the context of paratransit service, we think that the two minimum functional utilitarian aspects that need to be considered are the aspects of 'how paratransit can deliver passenger to destination' and 'safety' because both are the core benefits of transportation service.

The Hedonic Model gives several theoretical contributions for the literature of land transportation passengers' behavioural intentions. First, this model leads practitioners and researchers towards the idea that consumers see public transport modes and private vehicles in the similar way. Factor like image that can make someone have favourable behavioural intention to private vehicle can also be used to explain public transport passengers. Previous researches thought that public transports compete among themselves (e.g. Wen et al. 2005).

Second, this model can be used to explain not only passengers' loyalty, but also their tendency to switch and become loyal to private vehicles although those behaviours are deemed irrational. Previous passengers' behavioural intention models (e.g. Jen and Hu 2003; Wen et al. 2005; Lai and Chen 2011) focused on the efforts to explain behavioural intention as rational decision making process and did not take into considerations the emotional attachment that made the decision-making process became irrational (Falk et al. 2010; Solomon 2012; Earl 2012).

Third, this model is the first in the context of land public transport that took into account a hedonic variable, image, as a precursor of behavioural intention. Previous researches generally saw the functional utilitarian variable, service quality, as the precursor (e.g. Wen et al. 2005; Lai and Chen 2011). This idea might be the reason why efforts in improving service quality did not automatically improve passengers' behavioural intention. In the context of people with hedonic experiential needs, image holds an important role in forming passengers' consumption pattern, including their assessment on service quality (Falk et al. 2010; Wilcox et al. 2011; Solomon 2012; Earl 2012).

Fourth, Hedonic Model put image as switching barrier, preventing passengers to switch to other alternatives. Based on self-completion theory, image that match with passengers' identity and lifestyle would create emotional bound between the service and the passengers. An emotional bound will prevent them to leave. This is different with previous study that saw switching cost as the barrier (e.g. Wen et al. 2005). The use of image instead of switching cost as switching barrier will help us to distinguish loyal customers with the ones that consumed out of 'lack of alternatives' and customers that did not have the resource to use other alternatives. Furthermore, the use of image as switching barrier will strengthen the argument that loyalty can make customers insensitive towards price and the existence of competitors (Jain and Singhal 2012).

Fifth, Hedonic Model reveals that perceived value is a comparison between images and perceived sacrifice. In accordance with the view of this study that the means of transport are related to hedonic experiential needs, this model holds that the perceived benefit in the mind of consumers are the ones related to the fulfilment of hedonic experiential needs rather than functional utilitarian aspects. This is different with previous models that emphasized on service quality which represents the fulfilment of functional utilitarian needs (Wen et al. 2005; Lai and Chen 2011). Nevertheless, as previously described, it does not mean that service quality is to be ignored because people with hedonic experiential needs still consider minimum functional utilitarian needs at some point (Falk et al. 2010), which are the core benefit of the product (see Kotler and Keller (2012) for a discussion of product classification).

6.2 Managerial Implications

The study found Hedonic Model to explain paratransit passengers' behavioural intention. Referring to this model, paratransit passengers' behavioural intention is directly and positively influenced by perceived value and image. In this context, perceived value serves to attract passengers while the image provides switching barrier, making it very unlikely for passengers to switch to other alternatives, including the private vehicle.

Perceived value of paratransit service in the eyes of the passengers is determined from the ratio of perceived image and perceived sacrifice. Furthermore, in the context of hedonic needs-based consumption, customers will perceive image as a benefit when the minimum functional utilitarian requirements, core benefits, have been met (Falk et al. 2010) while the performance of the 'extra' functional utilitarian aspects of services will be biased because it is influenced by perceived image (Wilcox et al. 2011; Solomon 2012).

Hedonic model is built on the idea that customers perceive the roles and functions of public transport and other means of transport, especially private vehicles, similarly. This leads to a model developed based on the general phenomenon found on the passengers of land means of transport. Therefore, we expect that the implications will affect not only practitioners and entrepreneurs in the field of paratransit, but also practitioners on other public land transport in general.

Based on the above findings, in general, in order to create favourable behavioural intention, managers should be able to (1) present positive image of public transport service, (2) provide a high-perceived value for passengers, and (3) create minimum perceived sacrifice. They also must ensure that public transport services meet the minimum functional utilitarian requirements, the core benefits, such as 'drive up to the destination' and 'assure passengers' safety'.

Positive image of public transportation will arise if the service provider can manage it certain ways that make public transport stands in harmony with passengers' identity and lifestyle (Solomon 2012). In other words, public transport should be perceived as an illustration of one's self. A matched image will create strong emotional attachment (Tuskej et al. 2013) which can results in affective commitment and at the end leads to the lack of low cost-benefit calculations (calculative commitment) and a high sense of guilt when they consider to switch to other means of transport (normative commitment) (Kelly 2004). Furthermore, a person who has a strong emotional bond with the public transport will become 'zealouts', someone outside service providers who becomes leading advocates for public transport (Aggarwal 2004).

In the context of this study, the descriptive statistics (see Table 5.2) show that public transportation passengers are quite likely to give positive evaluation of perceived value and perceived sacrifice, but negative on image. This condition indicates that service providers have difficulties to prevent (lock) passengers from switching when there are objectively better options. Given this, service providers need to immediately fix the image of public transport so it can be aligned with the identity and lifestyle of the passengers.

6.3 Policy Implications

In Indonesia, the provision of public transport is a form of public service which is the responsibility of government. Organization of public transport in Indonesia is regulated by Act No. 38/2004 on Roads; Act no. 22/2009 concerning LLAJ (Road Traffic & Transport); Decree of the Minister of Transportation No. KM 49/2005 concerning Sistranas (National Transportation Systems), Government Regulation No. 8 of 2011 on Multimodal Transport; Regulation of the Minister of Transportation No. 8 Year 2012 on the Performance and Management Multimodal Transport; Decree of the Minister of Transportation No KM 35/2003 on the execution of public transportation.

The government regulates many aspects related to public transport, such as routes, fleet size, minimum service standards and tariff. Government policies related to these aspects led on some implications on the performance of public transportation especially the ones that correlate with variables identified as having significant influence on behavioural intention. Given that, the results of this study also provide some policy implications for the government in an effort to achieve favourable behavioural intention. Furthermore, in the context of local government at Jabodetabek, it is more essential given the importance to address the increasingly acute congestion. Here are some relevant policy implications of this research.

6.3.1 *Managing the Identity and Lifestyle of the Society of Jabodetabek*

The results of this study indicate that image has the largest contribution in the formation of passengers' behavioural intention. Referring to the self-completion theory, the image of public transport in the eyes of the passengers will be very positive and they will have emotional bond if public transport can symbolize someone's identity and even become their ways of life (Solomon 2012).

One of the distinguishing characteristics of public transport with private vehicle is the fact that public transport cannot be owned. In other words, passengers look at public transport as public property. This condition will cause a problem if 'identity' and 'lifestyle' mean owning private vehicles.

Indonesians are known as a nation that emphasizes communal and shared values (Tripoli 2011). This is also reflected in the national ideology, Pancasila. Nevertheless, as the era of globalization where information can be exchanged easily, values that emphasize communalism began to disappear and to change into materialism. People began to think that individual possessions can boost their status (Koentjaraningrat 2004). In this context, it will be difficult for public transport to build an image that can lead to emotional bond.

The government does not only act as servant to the citizens but also guide of its citizens so they can have appropriate and compatible ideology with the one espoused by the State (Janda et al. 2009). Furthermore, in the context of Indonesia, government has a role to influence people's identity and lifestyle. Given this, the government needs to manage the identity and lifestyle of the people so as not to emphasize the ownership of private vehicles as status symbol. This can be done through formal and non-formal education such as schools.

6.3.2 Improving Paratransit Operator Settings

The findings of this study suggest that in order to foster favourable behavioural intention, public transport operators should be able to give positive image and a valuable service. To achieve this goal, the performance of service providers needs to be improved in order to be equal with automotive manufacturers, providers of private vehicles.

Decree of the Minister of Transportation No. KM 35/2005 determines that public transport can be provided by (1) State-Owned Enterprise or Regional-Owned Enterprises, (2) national private companies, (3) cooperative and (4) individual citizens of Indonesia. In the context of Jabodetabek, individual owners are more common. This led to many numbers of service providers that do not have good management systems, or even part of decent financial management, marketing, servicing and maintenance. Consequently, performances provided for the passengers are not good enough.

Government needs to regulate the ownership so that public transports are owned and managed by legal entities such as: cooperatives, national private-owned enterprises, state-owned enterprises and/or regional-owned enterprises. Management by legal entities is expected to bring about better management system. Furthermore, good management system is expected to produce work programmes that can lead to valuable public transportation services with positive image.

Nevertheless, considering the current number of private ownership and re-organization will require considerable expense, individual ownership can still be allowed under fix management scheme provided by legal entity. Proposed models for the implementation can be seen in Fig. 6.2.

The model shows there are five parties with distinct roles in providing the service, which are the government, investor (owner of the vehicle), the company, the drivers and conductors, as well as the public. In general, the provision of public

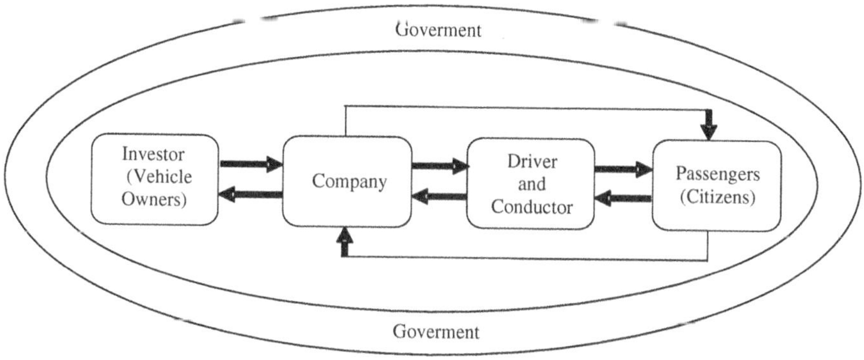

Fig. 6.2 Model for the provision of public transportation in Jabodetabek

transport is the responsibility of the government as mandated by Act No. 22 Year 2009 concerning LLAJ (Road Traffic and Transport). The government is responsible for coordinating all parties involved in the execution of public transport services. The relationship between all the parties can be explained as follows:

1. Investor Relations (Vehicle Owners) with the Company

The model shows that the provision of public transport service is directly conducted by the company and the investor is only responsible for the delivery of capital. In other words, the company is service operator and investors are only capital owners. Companies can take form of cooperatives, national private companies, state-owned or regional-owned enterprises. The obligation of the investors is limited to providing capital or vehicles while the return would be in the form of money from corporate profits agreed between the two sides. The system results in owners releasing their right as service providers, but they still can supply vehicles for companies providing transportation services. Thus, the provisions of public transport by the company are expected to be more organized with services that focus on the customer.

2. Company's relationship with driver and conductor

The model shows that drivers and conductors will not have direct relationship with the individual owners of vehicles. Every driver and conductor will be under company responsibility. They must be registered with one of the companies. The companies are obliged to provide education, training, career development and the driver's salary/wages. Thus, it is expected that the drivers and conductors will work more professionally.

3. The driver and conductor relationship with the passengers

The drivers and conductors are in the frontline. Principles of service excellence should be the basis for them in providing services. The drivers and conductors work on behalf of the company. Passengers should pay on appropriate rates and use public transport in a manner specified in the rules.

4. The company's relationship with the society

The model shows that raising awareness to use public transport falls under company's job description. Furthermore, companies need to socialize and communicate the importance of using public transportation. Passengers can aim their suggestions, criticisms and complaints to the companies to improve the performance of public transportation services.

6.3.3 Increasing the Perceived Value Using Integrated Public Services Policy

The results of this study indicate that perceived value affects passengers' behavioral intention city transport passengers. Within this corridor, the Government can take a major role to improve the perception of the benefits perceived by society when using public transport.

Currently, local governments in Jabodetabek area are competing to provide free public services to the society, particularly in education and health. Local governments can issue a policy that integrates free public service and public transport services in one package. In other words, free public services will only be given to people who use public transport. This condition is expected to heighten perceived value or benefits.

In addition, perceived value can also be increased by decreasing perceived sacrifice. Local governments have the authority to set the tariff. They can establish a payment mechanism by consumers based on 'route' and in form of a package for all types of public transportation services. For example, let us say, someone who would like to reach point B from point A needs to climb the three public transports. Pricing should be bundle, done only once at a cheaper cost. Furthermore, the payment should be done flexibly. Technically, this can be done by using tools such as communication reload cards.

References

Act No. 22 Year 2009 Concerning Road Traffic and Transport (In Indonesia language: Undang–Undang (UU) No. 22 Tahun 2009 tentang Lalu Lintas dan Angkutan Jalan).

Act No. 38 Year 2004 on Roads (In Indonesia language : Undang–Undang (UU) No. 38 Tahun 2004 tentang Jalan).

Aggarwal, P. (2004). The effects of brand relationship norms on consumer attitudes and behavior. *Journal of Consumer Research*, *31*, 87–101.

Decree of the Minister of Transportation No KM 35 Year 2003 on the execution of public transportation (In Indonesia language: Keputusan Menteri Perhubungan No. KM. 35 Tahun 2003 tentang Penyelenggaraaan angkutan orang di jalan dengan kendaraan umum).

Decree of the Minister of Transportation No KM 49 Year 2005 Concerning Sistranas (National Transportation Systems) (In Indonesia language: Kepmenhub No. 49 Tahun 2005 tentang Sistranas (Sistem Transportasi Nasional)).
Earl, E. (2012). Experiential analysis of automotive consumption. *Journal of Business Research, 65*, 1067–1072.
Falk, T., Hammerschmidt, M., & Schepers, J. J. L. (2010). The service quality satisfaction link revisited: exploring a symmetries and dynamics. *Journal of the Academy of Marketing Science, 38*(3), 288–302.
Government Regulation No. 8 of 2011 on Multimodal Transport (In Indonesia language: Peraturan Pemerintah Nomor 8 Tahun 2011 tentang Angkutan Multimoda).
Jain, R., & Singhal, S. (2012). A conceptual model for exploring impact of loyalty programs on consumer purchase behavior: With special reference to Indian apparel retailers. *IRACST-International Journal of Research in Management and Technology*, 2(2), 164–173.
Janda, K., Berry, J. M., & Goldman, J. (2009). *The challenge of democracy american goverment in a global world*. Boston: Wadsworth.
Jen, W., & Hu, K. C. (2003). Application of perceived value model to identify factors affecting passengers' repurchases intention on city bus: A case of the Taipei metropolitan area. *Transportation, 30*, 307–327.
Kelly, S. J. (2004). Measuring attitudinal commitment in business-to-business channels. *Marketing Intelligence and Planning, 22*(6), 636–651.
Koentjaraningrat (2004), *Kebudayaan, Mentalitas, dan Pembangunan*. Jakarta: Gramedia.
Kotler, P., & Keller, K. L. (2012). *Marketing management* (14th ed.). England: Pearson.
Lai, W.-T., & Chen, C.-Fu. (2011). Behavioral intention of public transit passenger—the role of service quality, perceived value, satisfaction and involvement. *Transport Policy, 18*, 318–325.
Regulation of the Minister of Transportation No 8 Year 2012 on the Performance & Management Multimodal Transport (In Indonesia Languange : Peraturan Menteri Perhubungan Nomor 8 Tahun 2012 tentang Penyelenggaraan dan Pengusahaan Angkutan Multimoda).
Shimp, T. A. (2010). *Integrated marketing communication in advertising and promotion* (8th ed.). Cengage Learning: South-Western.
Solomon, M. R. (2012). *Consumer behavior: Buying, having and being* (10th ed.). London: Prentice Hall.
Tripoli, F. (2011). Kecerdasan ekonomi masyarakat indonesia belum berkembang. *Kompetisi*, Edition 28, 28.
Tuskej, U., Golob, U., & Podnar, K. (2013). The role of consumer-brand identification in building brand relationships. *Journal of Business Research, 66*(1), 53–59.
Wen, C.-H., Lan, L. W., & Cheng, H.-L. (2005). Structural equation modelling to determine passenger loyalty toward intercity bus services. *Journal of the Transportation Research Board, 1927*, 249–255.
Wilcox, K., Roggenveen, A. L., & Grewal, D. (2011). Shall I tell you now or later? Assimilation and contrast in the evaluation of experiential products. *Journal of Consumer Research, 38*(4), 763–773.

Chapter 7
Conclusion

Abstract This chapter is the final chapter of the book. It presents the conclusions, the limitations of the research and the next research agenda.

Keywods Conclusions · Research limitations · Research agenda

7.1 Conclusion

Human needs of mobility have brought significant growth on the number of vehicles. This then led to variety of environmental and social issues. Congestion, air pollution, noise pollution, dependence on fossil fuels and climate change are some of the negative effects. This condition is experienced by almost all countries in the world, including Indonesia, especially Jabodetabek area. Jabodetabek as a metropolitan area in Indonesia also have congestion and air pollution problems. Congestion in Jabodetabek area has caused economic losses from the sum of Rp. 3,000,000,000,000.00 for vehicle operating costs and Rp. 2,500,000,000,000.00 for travel time. In addition, the air quality in Jabodetabek area also decreases from year to year. In fact, Jabodetabek is one of the areas in the world with the worst air quality (JICA Study Team 2004).

The problem stems from the amount of vehicles operated in the road. More specifically, the problem is due to the huge number of private vehicles compared to public transport. The ratio between the two is 98 % for private vehicles and 2 % for public transportation (UTIC 2010). Ironically, private vehicles are only responsible for 44 % of total mobility. Meanwhile, the mere 2 % of public transportation facilitates 56 % of the total mobility (DTKJ 2012).

The declining number of public transportation passengers' indicates that people are not loyal to public transport. Unfortunately, the literature on public transport passengers' behavioural intentions, which is the desire to remain loyal and recommend it to others, is very hard to find. Therefore, it is necessary to study of

S. Sumaedi et al., *Public Transport Passengers' Behavioural Intentions*, SpringerBriefs in Business, DOI: 10.1007/978-981-4585-24-8_7,

passengers' behaviour. The study is crucial in finding factors that caused people prefer using private vehicles and leaving public transport. It is also important to find effective solutions on how to attract people to use public transport.

The research results show that the factors that influence paratransit passengers' behavioural intentions are perceived value, image and perceived sacrifice. Perceived value and image are the factors that directly affect the change of pararansit passengers' behavioural intentions. Furthermore, perceived sacrifice and image have indirect effects on paratransit passengers' behavioural intentions where perceived value acts as an intervening variable. Of the three factors, image is the one that has the greatest influence on paratransit passengers' behavioural intentions.

7.2 Research Limitations

The limitations of this study are as follows:

1. This research only studied land paratransit with small bus and passenger cars types in Jabodetabek. Consequently, the result of this study needs to be tested further for compliance with the phenomenon of passengers' behavioural intention of other kinds of land public transport.
2. The sampling technique used was convenience sampling. This is because there is no actual data related to passenger profiles as a base of research sampling techniques. In effect, sample ability to represent the whole population is considered low.
3. This type of research was based on cross-sectional data. Consequently, it is difficult to determine the time-series link across variables.
4. We did not involve other factors that may moderate the relationship among the constructs, such as the sociodemographic of the respondents.

7.3 Recommendations for Future Research

This study has provided a comprehensive framework for understanding the antecedents of paratransit passengers' behavioural intention in Jabodetabek. The factors used are perceived sacrifice, image, perceived service quality, perceived value, satisfaction and trust. The results of this study demonstrate that paratransit passengers' behavioural intention affected by perceived sacrifice, image and perceived value. Even so, the authors are also aware of the limitations of this study, so recommendations for future research are needed. First, further research should add other antecedent variables that possibly can affect behavioural intention because the value of coefficient of determination of our model is 48 %, so it is possible there are many other factors that influence paratransit passengers' behavioural intentions. Second, there is a discrepancy with the findings of previous studies, in which some other researchers said that perceived service quality, satisfaction and trust can influence behavioural intentions. Therefore, to fill the gap, more comprehensive studies are needed by adding moderating variables such as age, gender, income, culture, lifestyle, etc. The addition of these variables might provide more interesting results. Third, further research should employ longitudinal designs, making it possible to provide answers for the gaps between these studies with the previous ones. Fourth, further research needs to investigate ways to increase perceived value of public transportation services, since that factor has the greatest impact on paratransit passengers' behavioural intention.

References

DTKJ (Dewan Transportasi Kota Jakarta). (2012). *Ibukota yang tak punya rencana.* Retrieved from http://dtk-jakarta.or.id/v3/en/slide-view/ibukota-negara-yang-tak-punya-rencana/

JICA Study Team. (2004). *The study on integrated transportation master plan for Jabodetabek (Phase II), final report, main report, Vol. 1: Master plan study, in coordination with National Development Planning Agency (BAPPENAS) Republic Indonesia.*

UTIC (Urban Transport Information Center). (2010). *Transportasi Kota Jakarta mengkhawatirkan.* Retrieved from http://bstp.hubdat.web.id/index.php?mod=detilSorotan&idMenuKiri=345&idSorotan=54

Index

S. Sumaedi et al., *Public Transport Passengers' Behavioural Intentions*, SpringerBriefs in Business, DOI: 10.1007/978-981-4585-24-8,

The manufacturer's authorised representative in the EU is Springer Nature Customer Service Centre GmbH, Europaplatz 3, 69115 Heidelberg, Germany. If you have any concerns regarding our products, please contact ProductSafety@springernature.com

Printed and bound by CPI Group (UK) Ltd, Croydon, CR0 4YY
30/09/2026
02225258-0004